# Workers at Risk

# WORKERS AT RISK

## Voices from the Workplace

## Dorothy Nelkin
## Michael S. Brown

The University of Chicago Press
Chicago and London

DOROTHY NELKIN has written extensively on controversial issues of public policy in science and technology. She is a professor in the Department of Sociology and in the Program on Science, Technology, and Society, at Cornell University. MICHAEL S. BROWN is a research support specialist and a Ph.D. student in the Program on Science, Technology, and Society, at Cornell University.

The University of Chicago Press, Chicago 60637
The University of Chicago Press, Ltd., London

©1984 by The University of Chicago
All rights reserved. Published 1984
Printed in the United States of America
91 90 89 88 87 86 85 84    5 4 3 2 1

**Library of Congress Cataloging in Publication Data**

Nelkin, Dorothy.
  Workers at risk.

  Bibliography: p.
  Includes index.
    1. Industrial hygiene—United States.  2. Industrial safety—United States.  3. Labor and Laboring classes—United States—Attitudes.  4. Chemicals—Hygienic aspects.
  I. Brown, Michael Stuart.  II. Title.
  HD7654.N45  1984      363.1'1'0973      83-9319
  ISBN 0-226-57127-0

# Contents

"Controlling Our Lives"
The Workplace
Technological Change
Political Concerns

# Preface

This book developed out of our concern that the burgeoning field of risk assessment was neglecting a key aspect of the problem of occupational health—the perceptions and concerns of workers themselves. In an effort to understand these perceptions and the social factors that influence them, we sought a method of research that would capture the subtleties of the environment of work and the realities of economic choice as they affect workers' views of risk. Our approach follows a tradition of qualitative research that focuses on the meaning and social context of human behavior, the importance of subjective experience, and the connections between such experience and behavior.

The main part of our research consisted of extended and open-ended interviews with people who work with chemicals in a wide range of occupations. We contacted people through international unions, health and safety classes, conferences, Committees on Safety and Health (COSH), and through casual conversations. After a number of preliminary trials, we conducted, taped, and transcribed interviews with 75 workers representing a variety of occupations. Twenty-nine work as chemical operators, maintenance workers, and technicians in large chemical, pharmaceutical, food processing, or industrial plants. Four are maintenance or technical workers in small production firms. We talked to 16 laboratory technicians and maintenance workers in universities, agencies, or research institutes, four fire fighters, four railroad workers, three health care workers, four artists, and three gardeners. We also interviewed a variety of individuals, including a beautician, a dry cleaner, a deckhand, some painters, a woodworker, and several clerical workers.

Among our 75 respondents, 44 belong to unions. Eight are under 24 years old, 57 between 25–44, eight between 45–64, and two are over 64. Twenty-eight are women. Thirty-four of the 75 people we interviewed identified themselves as health and safety activists. For further biographical details on individuals and their jobs, see Appendix 1.

Beyond our interviews with workers, we also talked to people in industry, federal and state agencies, universities, and unions. These included company doctors and hygienists, corporate safety directors, international union representatives, government regulators, scientific specialists in occupational health, and risk analysts. Because of our focus on workers, we did not excerpt selections from these interviews, but rather used them to provide background and perspective as we prepared the interviews with workers and later analyzed them.

These interviews with workers were loosely structured and flexible. The questions varied according to the occupation and work setting of the person interviewed. To preserve the style of individuals and the context of their responses, we recorded and transcribed the interviews and analyzed the more than 3,000 pages of transcript to determine predominant themes. We then chose excerpts that best characterized the views of our respondents, edited them for clarity, and organized them according to these themes. For background to the issues raised in these excerpts we have provided extensive bibliographic material.

Many people worked on this project. Lisa Norling, Joan Parker, and John Wooding helped with interviews and contributed to shaping the project. Joan Parker was invaluable in helping to set up interviews. Lisa Norling used impressive administrative skills to hold the project together. Discussions with Steve Cupery, Tony Mazzocchi, Mark Nelson, Marilyn Powers, Mark Sagoff, Margaret Seminario, and Vickie Stone helped to frame the issues and keep us informed about current political and policy issues. For detailed reading and critical comments on the manuscript during various stages of writing we are indebted to Alice Cook, Mary Douglas, William Friedland, June Fessenden-Raden, Frank Goldsmith, Steve Hilgartner, Sheila Jasanoff, Laura Malakoff, Walter Malakoff, W. Curtiss Priest, Paul Slovic, and William Foote Whyte.

We are particularly indebted to the many workers who allowed us to interview them. Their patient and informative responses are the essence of this book. We promised them anonymity, and we have not used their real names.

Appendix 1 provides brief biographical sketches of those workers whose voices are recorded in the book, and the pages on which they appear. This Appendix and the detailed Table of Contents serve the purpose of a general index.

This material is based on work supported by the National Science Foundation Program on Ethics and Values in Science and Technology, under grant no. ISP 8112920. Any opinions, findings, and conclusions expressed here are those of the authors and do not necessarily reflect the views of the National Science Foundation.

# Introduction
## The Dangerous Trades

In 1910 Alice Hamilton, one of the first American specialists in the field of occupational disease, began to study the ravages of the "dangerous trades." She was staggered by the problems she found among felt hat and lead workers, and remarked that knowledge about the health effects of new chemicals depended on the use of workers as "guinea pigs." Today, more than 70 years later, the dangers of working with toxic substances are a focus of growing public attention. We hear of asbestos workers with lung cancer and vinyl chloride workers with liver cancer. We hear of neurological disorders among pesticide workers and sterility problems among pharmaceutical workers. The public is only beginning to come to grips with the political, legal, and ethical implications of occupational risks.

In this book we look at the experiences of the men and women who work with toxic chemicals. How do they perceive the dangers of their jobs? What do they feel about working with substances that may affect their health? How do they cope with risks? Through such questions we explore several dimensions of the problem of occupational disease: the anxieties and concerns of those who face hazards in the workplace, their attitudes and adaptations as they cope with conflicts between the risks and the benefits of their work, and their actions in response to risk.

Much of the literature on risk analysis assumes that risks can be objectively measured by estimating the extent of exposure and the probability of accident or disease. Evaluating risk is viewed primarily as a problem of measurement, based on the judgment of scientific and medical experts. Some analysts supplement these measurements with calculations of the characteristics of different risks and their bearing on acceptability. People tend to underestimate the dangers of familiar risks such as driving a car, and dismiss the hazards involved in voluntary activities such as sports. They find it most difficult to deal with risks that are involuntary, unfamiliar, ill-defined, and potentially catastrophic.

These approaches focus on the nature of the hazards themselves. An alternative approach views risk as embedded in a broader context of social and political relationships. This approach, identified with the work of anthropologist Mary Douglas, sees definitions of risk as comments on the social order, inspired by visions of justice and filtered through values, interests, and goals. It suggests that risk is a social concept and that perceptions depend as much on the social context as on the nature of the hazards themselves.

Our research assumes that perceptions of risk and adaptations to hazardous work take place in a nexus of social, political, and economic conditions. We investigate how such conditions influence attitudes and responses toward risk by focusing on people who are exposed to toxic chemicals in the course of their daily work. We have chosen the problem of chemical hazards for several reasons. Their ubiquity in so many occupations suggests the widespread importance of the problem. But also, the uncertainties involved in the objective evaluation of risks allow for a variety of interpretations as to the nature and extent of workplace hazards. This opens possibilities to assess the impact of social factors across a wide range of occupations.

Chemical products are indispensable to the modern industrial economy. Industry uses an estimated 63,000 commercial chemicals (with hundreds more being added annually) to create products including basic chemicals, pharmaceuticals, plastics, paints, and pesticides. The ubiquity of chemicals in the environment is a problem, not only for those who work in the chemical industry but for all of us as we clean our ovens, paint our houses, spray our gardens, and polish our boots. We accept their presence in our everyday lives. Workers, however, are especially vulnerable. Those potentially exposed to these products include the 4.6 million employees in chemical and chemical products industries, along with others who use chemical products in occupations as varied as fire-fighting, glass making, fine art, gardening, scientific research, food production, nursing, and maintenance work.

The cost of our dependence on chemicals in terms of the health of the people who make and use chemical products has become increasingly evident over the past 20 years.[1] The health effects of exposure to some substances are well understood, for example, ammonia causes severe eye and respiratory problems, vinyl chloride can cause a variety of cancers, and lead can cause neurological damage. However, in the case of most chemicals, uncertainty prevails. These uncertainties are compounded by the difficulties of evaluation. The effects of chronic chemical exposure become evident only after a latency period of many years. The cumulative impact of prolonged exposure to low doses, as well as the effects of exposure to combinations of substances, confound systematic analysis. Experiments with laboratory animals are time-consuming and expensive. Poor data collection, relatively primitive diagnostic techniques, and the shortage of physicians trained in occupational health have obstructed the accumulation of information over time.

The uncertainty of risk analysis—the fact that the data do not provide unique answers—allows for a range of perceptions about the nature of chemical hazards and their effects on worker health. Disparities in perception contribute to disputes over the health of workers—disputes that reflect the conflicting values that guide evaluations of risk. Some place concerns about risk within a calculus of economic viability; others see risk from the point of view of those exposed to hazards. Each view has implications for responsibility and control.

Understanding this conflict in its multiple perspectives is critical. The assumptions underlying industrial practices and policies are well articulated in the economic and policy literature (see Bibliographic Essay). Yet relatively few studies have focused on workers' perceptions of risks. Worker awareness of health hazards was minimal before the mid-1970s, but perceptions began to change following the creation of the Occupational Safety and Health Administration (OSHA) in 1970 and the implementation of its policies over the next decade. Mass media coverage of several dramatic occupational health incidents increased worker awareness. The appointment in 1977 of Eula Bingham as OSHA assistant secretary, and the new emphasis on prevention of occupational illness, further highlighted the issue for workers.

These changing attitudes were reflected in the findings of a series of national Quality of Employment surveys conducted by the University of Michigan's Institute for Social Research in 1969, 1972, and 1977. The surveys probed the opinions of workers about their work lives, and included questions about occupational health and safety. The responses suggest the growing awareness and increasing concern about health among

1. The U.S. Department of Labor estimated that in 1978 nearly two million workers suffered from occupational diseases, and the total income lost from their illnesses came to $11.4 billion. Because of the difficulty of obtaining accurate data, estimates of the prevalence of occupational diseases vary, often according to the interest of the reporting organization.

many production workers. They reported a remarkable increase in worker sensitivity to a range of work-related hazards. Between 1969 and 1977, those who said they were exposed to one or more hazards on the job increased from 38 percent to 78 percent. In particular, workers were increasingly sensitive to health-type problems as contrasted to traumatic injuries.

The Quality of Employment surveys point to significant changes in worker's perceptions of risk, but tell us little about the differences among them and the conditions that influence their concerns. While providing statistical evidence of changing risk perceptions, the surveys do not convey what people know about hazards at work, how they experience them, and how they cope.

We chose to complement this survey research by exploring the variety and depth of workers' experiences rather than the generalities of statistical aggregates. In our interviews with workers we did not try to evaluate the extent of their exposure or their actual state of health. Nor did we try to assess the validity of their perceptions of dangers (there are, indeed, a number of cases where perceptions are based on erroneous understanding of toxic effects).[2] We have assumed rather that the perceptions, feelings, and desires of workers are an important reality in and of themselves.

As workers described their concerns, they were often negative. People tend to focus on problems because unproblematic experiences are less noteworthy. By presenting their experiences, we are not suggesting that they represent the entire spectrum of conditions among the millions of workers in the industries and trades represented. Rather we tried to understand the factors influencing their different responses to perceived risk.

To identify the social considerations that influence these responses, we interviewed people in diverse occupations and different work environments, including artists, railroad workers, chemical workers, nurses, fire fighters, pharmaceutical workers, laboratory technicians, and gardeners (see Appendix 1). We encouraged them to talk freely about (1) the nature of their jobs, the supervisory arrangements, the general conditions of the workplace, and their attitudes toward their work; (2) their personal experiences, observations, and anxieties about health; (3) their ways of coping with or adapting to the problems they have confronted or observed; (4) their views about responsibility for conditions in the workplace and the government's role in setting and enforcing standards; (5) their knowledge about hazardous substances, their use of information, and what they want to know; and (6) their personal background, such as

---

2. See Appendixes 2A and 2B for a summary of technical information about chemicals identified as a problem by respondents.

their family obligations, job mobility, and general views about technology and politics.

We present the views of workers in their own words. They are subjective statements lacking the dispassionate objective tone of scientific discourse, but we believe that the perceptions, anxieties, and behavior of workers cannot be dismissed. To discount them would be to oversimplify the complexities of hazardous situations and to deny the value of experience as relevant data in the understanding and reduction of risk.

# PART 1

# Jobs and Risks

# 1  Working with Chemicals

What is it like to work in a modern chemical factory, a hospital, a railroad, or a freighter? What do people do every day on their jobs? How do they feel about their work and its risks? The people in our study include blue-collar and service workers, small business owners and employees, professionals and artists. What they share is regular exposure to toxic chemicals as a part of their jobs.

New technologies and different processes have altered what people do in their work lives and the conditions under which they labor. Certainly, conditions in the modern chemical factory have changed dramatically since Alice Hamilton described the "dangerous trades." There has been an explosion in productive capacity and new substances, along with changes in production techniques. Chemicals used to be produced primarily in batches; workers would dump raw materials into open vats to be mixed, empty the vats, and package the product. Today many modern companies use continuous flow methods. Production takes place within enclosed vessels involving fewer workers using less brawn. They work as control room operators, taking readings of temperatures, pressures, and rates of flow.

This "process technology" does not necessarily mean that the work is antiseptically clean. Leaks still occur, spilling raw materials and products. Batch processing still exists for many specialty chemicals. As our respondents suggest, maintaining equipment and cleaning filters is still dirty work. And the problem of workplace exposure hardly ends once the product leaves the

3

chemical plant because chemical products are used throughout industry. According to the University of Michigan's Quality of Employment survey of 1977, 72 percent of male production workers reported exposure to air pollution and fumes, 45 percent to dangerous chemicals. Combining all professions, the survey found 39.7 percent of workers reporting exposure to air pollution and 28.9 percent to dangerous chemicals.

Along with chemical production workers, we interviewed people who work with toxic substances in many other occupations: beauticians who use chemicals to dye and alter the shape of hair; deckhands who "bust rust" and repaint their ships; computer assemblers who dip components into solvents to clean them; artists who use paints, resins, and solvents; and fire fighters who confront combustion products that include the exotic gases given off by burning plastics.

In large enterprises, the extent of exposure to toxic substances is related to social status. It is mainly production and maintenance workers who are subject to risks. The managers of chemical or manufacturing plants, hospitals, museums, or railroads are rarely exposed on a regular basis to the substances used and produced in their plants. These differences in exposure to hazards reinforce the distance between upper management and production workers, and raise difficult questions of distributive justice.

The situation is often different for the self-employed. The owner of a small beauty shop toils along with the other hairdressers; the artist working alone takes his or her own risks. These people play a double role as managers and workers, making decisions as well as bearing the consequences as they directly engage in their work.

We seldom glimpse the mundane details of other people's jobs. We have little idea of the routine and the risks that are involved in producing what we use every day. In this chapter, workers in a variety of occupations describe their work and its risks.

---

### Walter, pipe fitter, glass factory

The company makes glass containers, small beverage containers, catsup, mayonnaise, whiskey bottles, salad dressing, Ragu sauce, most any container. One of the mainstays is beer bottles. They make a lot of them. One machine makes 300 bottles a minute running 24 hours a day. Of course some of them are bad and some good; the bad ones are trucked back over the litter belt or through the cullet system. Cullet is used glass that's no good and they crush it and then put it back in the batch and remelt it. It makes a stronger glass.

We use lots of toxic chemicals; for example, titanium tetrachloride is used with a fluorocarbon. These two are combined, put under pressure, and sprayed on the outside of a bottle for hot-end coating. Then we bake it in. When the hoods start to plug up, the fumes come out around. We also use silica sand, soda ash, sodium sulfate, and cyanide. This is just some of the stuff that goes into making glass. Probably 10 gallons of chlorine like they use in swimming pools everyday goes into our shear spray system. Then we have algaecides with different numbers on them. It tells right on them, "Don't get it on you." It can cause sores, ulcers, and all this junk. Sulfur flour is used to clean the molds. We dip a pipe into a bucket of sulfur flour and take an air gun and blow it into the molds. It burns the carbon off. It stinks. In fact I have a little asthma problem and if I get direct contact with the sulfur, that really tears me up. Some of this stuff I know gets dumped down a drain now and then. Not purposely. It's just an accident when it happens. They try to be careful because they don't want the fines. Then we have a recycling water system where they cool the glass. All the stuff from the spray system is water soluble, so the water is white. If you fell into it, it would kill you. An engineer told me that at one of the other plants that uses the same stuff— a dog fell into the pond. They fished the dog out and he went over and laid down. The next morning they came back and he still laid there— dead. It killed him.

### Fred, chemical operator, chemical plant

My main job is running the columns, but I assist the other operators if they have anything that's not going right. So I work all over the place doing minor maintenance.

Each day I spend about five minutes talking with the man I relieve. We go over any problems he had before me and that I might run into on my shift. Then I go over to the plant and check my equipment— especially after the midnight guy. He's tired; he might not be alert, so I make sure that everything is right. I check the storage tank to make sure that water doesn't destroy the product. Then I make out forms on any samples that have been taken to get ready for my shift. Every hour you have to get readings on the refrigerators and make sure they are running properly and that the pumps are not overheated.

A lot of times we get problems when we first take over the equipment. The guys relax a little when they are intending to go off so it's better to check things over for safety when you first come on. You might have a small HCl (hydrogen chloride) leak on a reactor and you help the guy work on it. You might have a bad diaphragm on a valve and you put a vacuum on it. The supervisors rely on the chief to help them run the ship properly. We get a little more money and we have a little more responsibility.

Most of the time things work and we're lucky. But if you think about how bad it is, you better get out. We've had leaks where the stuff sprayed so fine that you could walk right into it and absorb it into your mucous membranes and your eyes. I had a big accident once when we started up the first charge after a vacation. During the week I was off, they had cleaned the screens on the mix tanks where they agitate the cyanide and alcohol. That mixes them thoroughly and they add this at a controlled rate to the reactors. The cyanide reacts with the iron and forms fine black stuff that will foul up the pump. The screens collect most of it. I didn't know they had cleaned the screens and I set up the valves and opened up everything, so the cyanide that was trapped in the line drained right into the tank. I ran over, holding my breath. Because the valve had been left open for several weeks, it was stiff. I couldn't close it. The other guy was coughing and gagging from the cyanide—he got a pretty good whiff of it. My head was spinning. I ran to the fan because I couldn't seem to breathe right and that's the last I remember. The guys told me I fell into the fan and slid to the floor. They took me outside and laid me on the ground and got the oxygen. I came to after about ten minutes. The boss came down and said, "You'll have to go to the hospital." I didn't want to go but they made me, which was right. They kept me there 48 hours, gave me a cardiogram, blood tests, chest X-ray. We've had similar things happen to at least half a dozen guys who were exposed to cyanide from leaks.

### *Laura, filter cleaner, pharmaceutical plant*

I work with all different kinds of filters; some of them are fiberglass, some I don't really know what they are. I usually work on the roof. I shut down the air supply units and pull the collectors out, and put the new ones in. I get filthy dirty—it's not much of a job, as far as I'm concerned. I work on the south end, because there's areas on the north end where they ban women of child-bearing age because of certain drugs that they're working on.

There's certain things that bother me. They have filters, called "absolute filters," that are like 98 percent perfect. Nothing gets through them. They use those in labs with hoods where they're working with real super toxics, so that nothing is taken out of the hood or put into the hood that would affect the experiment. Unfortunately, they forget to tell us what's in there. One day I went into a lab and saw a sign on the hood saying SUPER TOXICS. We had come in to change the filter and they never let us know. The first time a bag broke in that one building, I went up and got real dizzy. I didn't have a mask on. They never told me I should wear a respirator before going on that job. It really ticked me off.

Sometimes the filters you come into contact with have just regular old dirt, sometimes they have product in them. The filters, depending on where they are, get dirty in a week. Some of them you change twice a year, some, many many times. If you're cleaning a filter and you make a mistake, you could really cause some damage to the product or the experiment they're working on. If you screw up and don't change any of the filters, say, in the toxicology building 'til it got to a really serious point, then you might have problems with the temperature control. All the test labs are temperature controlled and you could lose animals and then you'd lose your job along with the animals.

*Arnie, chemical operator, food processing plant*

As a bulking operator in a food processing plant, I was mixing the component oils, the spice oils, into the final mixture that would be sent out to a larger food processing firm. For example, if we were mixing up, oh, let's say vegetone, which is a food coloring, you might dilute that with oils and you might add certain antifoam additives or other things which would give it a heavier or a lighter consistency. It would then be put either into drums or smaller pails and be sent off to the manufacturers that use it. We were working with huge vats of maybe 20 to 30 thousand pounds.

Health and safety was a big issue in the plant because the conditions were just unbelievable. I mean it was like walking into a jungle. For example, we used to get these hot chili peppers from Uganda. We extracted the heat fractions out of these hot chili peppers and then used the color in paprika. When these chili peppers came from Uganda, they were so full of rat shit that they had what they called a shit sifter. It was this big vibrating screen. The shit was heavier than the dried chilis, so this huge big wire kind of funnel-shaped screen would vibrate and shake the entire batch of these hot chilis and the shit would fall through these screens. Well, the fuckin' chili dust was all over the place. The stuff would burn when we'd breathe it and I had chronic bronchitis. I had chronic bronchitis from the time I worked in there until the time I left. Every year I was in the hospital. I haven't been in a hospital since I quit. We finally got the shit sifter out of there. We called the FDA and told them that the company is getting stuff in from Uganda that's full of rat and goat shit, and they're selling it to people. They sent out this FDA inspector and this guy's looking every place but the obvious. Here's this huge big shit sifter full of hot chilis and he's walking around in the cleanest part of the plant. We dragged him in and threatened to report him, so he cited them and they had to eliminate the whole operation. They started to buy chilis from Red China and they were immaculate, I mean they were perfectly clean. It was funny, the stuff we got from China was just unbelievably clean. They probably hand-did everything, you know, because they all came in these handwoven baskets.

### Peter, railroad signal inspector

My job varies from day to day. Basically, I'm in construction; we do all the new construction for the railroad signals. It's mostly wire work, putting up relays. They always talk about our department as a jack-of-all-trades because we do everything. We have to know hydraulics because we have hydraulic gates. We have a lot of things that are electrical. We have pneumatic switches, so you have to know about that. We use mechanical interlockings which are about 100 years old, but amazingly they work. So things really vary. That's one of the good things about the job. But there are a lot of risks.

The relays and the wires have to be protected from each other by a nonflammable material, and they use asbestos. They only way to get to the wires is to drill holes into the asbestos. There are quite a few wires so it's quite a few holes.

I was also directly exposed to chemical spraying. We were painting the switch movements, and one night the track department came along and sprayed for grass and weeds, when I was right there on the crossing. They sprayed the interlockings and the switch mechanisms so we had to go out and wipe them off. It was all white. When we got there it looked like a white rain hit it, and we had to wipe all of them down with a rag and water. They used the same stuff when I was in communications. You get vines and things growing into the telephone lines, and they spray so they don't grow up again. They told us it's just a defoliant like you use at home. You get told all your life that everything is always safe, and it's checked by the government and that's who you believe. I only realized that there were problems at a union meeting when someone started speaking about what's happened to people exposed to it.

### Lee, stage carpenter, university

Backstage in the theater, we face all the hazards normally found in a shop situation—power tools for handling wood construction, power-driven staples, drill presses, saws with whirling blades in close proximity to human flesh. We use a cutting torch and arc welders that have been known—not in this shop, thank God—to weld soft contact lenses to people's corneas. We have to be constantly on guard. But the most noteworthy accidents these days are from the new methods of scenery construction. This used to be done with chicken wire and papier mâché. Now irregular shapes such as the bark of a large tree can be made more rapidly and realistically with pour-in-place polyurethane foam. The foam has to be mixed and poured and it expands and hardens rapidly. A gas escapes from the action of the foaming agent. One day a faculty designer was helping pour the foam and he collapsed. After a couple of hours he resumed work. Throughout the production of this play we were carrying him out of the room until he recovered, then bringing him back in again.

### Dorothy, deckhand

For about four years I worked as an ablebodied seaman on different kinds of boats, everything from charter sailboats and oceanographic vessels to small freighters, tramp freighters, and fishing boats. This was in the north Atlantic, mid-Atlantic, and the Caribbean. Most of my time was spent actually at sea, where your day-to-day routine involves standing watch. Depending on the system it would be either a 12-hour, a six-hour, or a four-hour watch. While you stand watch, there's different kinds of jobs you have to do, but very often the major task is maintenance.

The most common job is busting rust. That's for peons. Many of the ships are steel vessels, and, being at sea, they rust a lot. The job is to bust the rust and repaint the areas that are rusted. That sounds all very harmless but it's not. The rustbusting part of it is not so bad; you use all kinds of tools to scrape and beat on the metal. But the thing that bothered me about it is getting covered with metal dust all the time. Right after that, you do a whole lot of treatment before you paint. You use some sort of phosphoric acid; the trade name is OSPHO. After you prepare the surface of the metal with the acid solution, you put primers on. The primer that's used on every boat that I've ever worked on is red lead. It's real soft paint, when it's dry you can make a fingerprint in it. The reason it's so soft is because it has all this lead in it. I don't know if it's true, but some guy told me that red lead was outlawed in the United States: the EPA had banned it. He was one of the people responsible for buying supplies, so he said, "Well, we have to make sure that we arrange to always buy red lead when we're in a European or some other port, not in the United States."

Painting a ship above the water line, that's a piece of cake. You're out in the open air, and you just sand it down with this great big sander, paint a little of this, paint a little of that, and you're never really exposed to anything very seriously. But when you're painting this stuff in the hold below decks where there's no ventilation and you're squished into a little corner and it's dripping onto your face, that's the real problem. Or when you're hanging on the rigging on a sailboat with one hand and painting with the other, and the paint's falling all over you, you're constantly getting the stuff all over your skin. Or when you're taking off the old bottom paint, sanding it down with one of these big sanders, and the copper dust is flying everywhere. If you're doing bottom painting on a boat that's been partially hauled out for just a quickie maintenance job—

on some of the smaller boats, they just beach them and bottom paint them—that's especially bad because you're standing up to your knees or waist in the water. All the drips are falling into the water, and they form this oily scum of copper bottom paint on the top of the water, and it gets all over you. When you get out, your whole skin is just covered with this bottom paint. You have this sort of reddish or greenish tint for a week.

There's a lot of petroleum products on boats. All kinds of greases and oils and slimes of all descriptions. We use coal tar epoxy for water-proofing the inside of a hold, especially in the bilges of the ship underneath the lower floor plankings. That whole area is usually filled with water and pretty groady. We paint it with coal tar epoxy so that it doesn't rust so rapidly. If you're painting at sea, you usually have one person in the hold doing the painting, because that's all that will fit, and one person on top, sort of laying on deck looking into the hold to make sure the other person doesn't pass out. But you can't tell if they do, because the guy's just laying on his back anyway, and the only way you can tell if he's passed out is if his arm stops moving. So you hear people yell, "Are you just resting?!"

### Mike, photo lab processor, blueprint company

What do I do? We have a film processor; the machine has to be filled. There are three tanks, roughly five and a quarter gallons each, one for the developer and two for the fixer. I turn the machine on and fill the three tanks with cubes from Agfa, the company that supplies our chemicals. Basically, it's like any other job: repetitious, all day long. Once in a great while we'll have something that's hand developed, but it isn't too common. There's two kinds of processors, one is called a recirculator, which means it'll take certain chemicals in these cubes down below and recirculate the whole cube, which is great. The thing is it's a little bit dirtier system for processing film, because you're using the same chemicals over and over. It's like using dirty oil in a car. The other kind is a replenisher. Now this takes chemicals and keeping inducing them onto these trays up on top where the film goes through. So you always have live chemicals all over the place, a constant supply of fresh chemicals always getting in the air.

At one time we had to mix our own fix. We had to take the raw chemicals themselves, fill the tank with a certain amount of water, and mix in the chemical powders. You put your hand in and jiggle it around some, and your hand starts to, yecch, turn yellow, really yellow. Your nails turn yellow, your hands turn yellow. You have to stop working on these things for a while for your hands to turn back.

The chemicals are obviously not good for you. If they spill and you don't clean them off, what used to be a nice pair of pants will have a hole in them. And that's a fact. I have some home like that. They've been patched, patched, and repatched. The ones I got on I bought recently because, well, I had to. I don't wear them out, they just rust out, so to speak.

### James, computer assembly, manufacturing plant

I work with an inkjet printhead. It goes on bank machines and sprays the lettering on your checks when they're canceled. The previous ink was discontinued because when you washed your hands with soap it would break the ink down into benzene. Now we hear that the new ink is mutagenic, but it's considered a trade secret so we don't know for sure.

On the first floor of my building are the silicon chips. The second floor is a holding area for all the chemicals, pipes, and tanks. On the third floor where I work are printed circuits and copper boards. We use acids, trichloroethylene and all kinds of nasty stuff. The first floor has mostly perchloroethylene and a chemical called JP100, I'm not quite sure what that is. All I know is that the hazard ticket on it is pretty high compared even to perchlor. I worked with cupric acid for a while. It's a standard procedure to chew gum while you're working on the etcher so the cupric acid doesn't make your throat rough. Workers tell each other what to do and ways to get around the job, and one thing was chewing gum. This is what you have to do to keep things running smooth in your throat. It's ridiculous.

Before this job I spent two years in a warehouse area because I was a bad boy: I refused to work overtime. The thing that bothered me and a lot of other people was having to work 13 days straight and one day off, 10 hours a day. It got to be too much. You're exposed to a lot of chemicals. They always judge the exposure limit on an eight-hour day, and we were working a 10-hour day, 13 days straight. So we were getting a double dose. But what bothered me the most was the fact that we had no control. I wanted to go home and just recuperate for a while, that's all. So on Palm Sunday I just refused to go in. It caused a big uproar. A lot of people were backing me. I didn't get fired, but it didn't help my career.

They demoted me, and put me in an area tearing down old computers. There we worked with PCBs. We took the capacitators out of the old computers. They break in your hand and you get the fluid on you. We found out there was something bad with PCBs, so they made us wear gloves, and made sure our work clothes were washed at the plant, and that we didn't bring them home. But up until then, we were taking the capacitors out with screwdrivers, and when they'd break open we'd wipe it up with a rag. People didn't think it was that serious because you couldn't smell it. It was oozy, that's all.

*Ben, repairman, chemical plant*

I clean up equipment from different sections of the plant, especially the lime kiln. When equipment comes in, it's covered with this stuff that looks like dirt or rust. You don't know what you're handling. In order to clean it, I have to chip the stuff off, sand it off or blow it off.

There's a lot of overtime on the lime kiln because everything's always breaking down. It's just unbelievable what happens. You can't see, it's dark, the mud and lime dust is so thick, one breath and you're gagging. It's cold during the winter and it's warm during the summer. When you go down in the kiln to clean it, you've got to bundle up with hoods over the dust masks and a cheese cloth tied around your neck so you don't get burns. It's 115 or 120 degrees, and you have to go down there and shovel this stuff.

Another problem is the degreasing area. You dump the equipment down in a tank with trichloroethylene, let it degrease, and take it back out. Even though it evaporates fast in the air, you can still smell it. We put the equipment on straps and when we take the straps out they're covered with the chemical. You store them in your boxes. When I first opened up the tool boxes, the smell almost kicked my head back. Yet nobody knows too much about it and the average person wouldn't pay any attention.

Another part of the job is making Babbitt bearings. Babbitt looks like lead. It's a gray substance and you've got to melt it under heat, between 650 and 800 degrees. Within five minutes after you pour it out, it's hard as a rock. I don't know what's in it, and I've been unable to find out. I had them take a test to monitor me on the fumes that come out of it, and it was at the grade that is safe.

The plant is about 100 years old, and the equipment they've got—some of it is the original design. We do use an air impact wrench, instead of doing it by hand, but other than that there's no new technology that I see. In fact, in my opinion, if an outsider were to walk through that plant, he would have to think, this has got to be a setup. This has got to be something out of the movies. The things they have in there are unbelievable. Nothing's swept up, you're actually walking on years of product. They try to keep the main aisleways as clean as they can. In between, where you're working, it's dirty. If you were a manager and walked down through, you walk on a clean floor, but if you go off to the working areas, you couldn't walk through without coughing. If you were to walk in the regular eating area, I would put money on it that you would cough. If you were to go to the boiler house, you would taste the coal dust in your mouth, and you would come out with it on the skin of your face, except if you have safety glasses on. Then you'd be white around the eyes. If you were to walk in the soda ash, you would sneeze from the soda ash dust in the air. If you were to walk in the distillation building, there's an ammonia smell in there and you would gag. A new guy that starts here, his nose is running and his eyes are watery for the first month. Of course, you get used to it. You can take what's there everyday except when there's a leak.

### Lisa, laboratory technician, university

I work for a professor who is studying ion fluxes in nervous transmission, and my primary responsibility is to prepare the tissue that he uses for the experiments. This work involves cutting up electric eels every week and preparing their cells. There is a whole protocol. It takes about a day to prepare cells from the eel so that they can be used in the experiments that are being done by various people in the lab. In addition to preparing the cells, I also prepare chemicals, mix up buffer solutions, and clean up. The job was difficult for the first three months because my background in biochemistry is not very strong. There were a lot of simple things I should have known how to do but I had to learn: things like how many grams of sodium chloride you add to make an 8 percent solution. After three or four months, when I figured out what to do and became efficient at doing it, the job was no longer difficult.

We use toluene, and if there are leftovers of stuff that isn't radioactive it's just poured down the sink. For the first few months that I worked there, I would constantly ask where to dump things. It was always just, "Well, pour it down the sink." So I just poured everything down the sink. Everything. Eel remains, that was probably the least harmful. But also every kind of chemical that wasn't radioactive got poured down the sink.

### Rich, orchard worker

The job has different seasonal tasks: pruning the trees in the winter, harvesting the apples in the fall, and applying pesticides in the spring and summer. Apples take a huge amount of pesticides for the different diseases and insects that can attack them. So we used a wide range of pesticides, anywhere from organophosphates like Guthione to some of the fungicides, and different herbicides. Paraquat was one of them, also 2, 4, 5-T and 2, 4-D, before the ban.

In applying the pesticides, we use what we call the nurse-rig, which is a tank and a mixer, and a long hose with a pump coming off it. The person running the nurse-rig goes to a water tower, fills up the tank with water, adds the pesticides, and mixes it up while the other person is running the air-blast sprayer. Mixing the stuff up isn't difficult. Running the sprayer is a matter of timing. To be able to control your speed, check your trees, and make sure you're getting coverage, while maneuvering the sprayer, a huge machine, in and out of the orchards, takes a bit of skill.

There's a lot of pressure while the sprayer is running. You fill up the tank and get back to the sprayer before they finish. As soon as they start running out of spray you go full blast on the tractor to get the thing filled up as fast as possible so the guy can keep spraying. Wherever he stops, he gets filled up. That can be a problem. The person who ran the sprayer when I first got here had all sorts of emotional problems. Even when it wasn't necessary, he was constantly putting pressure on the rest of us running the nurse-rig. If we didn't move fast enough after we filled him up, he would start the spray up before we were out of there, so we got sprayed with pesticide.

### Tony, dry cleaner

I've been connected with dry cleaning for over 50 years. I got started because in 1929 there were no jobs. I got a job driving a truck for a dry cleaner, and the next thing I knew I was working inside. In two years I thought I knew all there was to know, so I opened my own business. Been at it ever since. It was a good time to start a business because everything had to be uphill. Even in the Depression, there was a limited amount of cleaning. Costs weren't high. Labor was cheap, supplies were cheap, and rent was cheap, so it didn't take too much to make a living.

I can tell you just a bit of the history of dry cleaning. Originally they started in with gasoline, and that was really treacherous. They cleaned things outdoors so they wouldn't burn the plant up, and hung things on lines to deodorize them. Then they developed a petroleum dry cleaning solvent which was much safer. It took quite a bit to get it burning. Once it did start burning, it was really a roaring fire, but it was not as hazardous as gasoline. In 1930 they came up with carbon tetrachloride which was a good cleaner, but pretty tough on the health. People who worked in it would get sick regularly. You were almost sure to have a stomach problem. Then they came up with an improvement called perchloroethylene. Perchloroethylene is pretty darn safe. You can be bothered by it a little bit, but you're not going to be poisoned or faint away. They changed to this after the war.

They also made new machinery. Originally dry cleaning machinery was like an open-air washing machine and the fumes were all over the place. They had big fans to blow them outdoors into the neighbors. Now the cost of perchloroethylene is so high, and there's the concern about health, so they save every bit of it. You just get a whiff now and then when somebody opens the door when they shouldn't. The fumes are sucked back into a large still and the still can recover 99 percent of the perchloroethylene you start with. It used to be wasted, it would just go out into the air. Now it's recovered, and, furthermore, cleaning plants don't smell like they used to. You used to be able to walk into a plant and tell right away that you were in one. Now you can walk up to the front entrance and not know it. It's not any different than a grocery store. I'll tell you, a worse smell is next door in the gas station. There's a repair garage there and they put something into the carburator to clean it, and it comes whooshing out of there and we get the smell of that. Much worse than cleaning fluid.

The chemicals that they used to use for spotting were straight chemicals, like alcohol and ammonia. Now chemical supply companies make special formulas: they put them in bottles marked for paint or for ink or for blood or for various things. If you have an ink spot, you put the ink remover on it. You don't have to be a chemist, but just a man who wants to work. Of course you develop a technique, but it's simple compared to what it used to be. Up to that time, the man who took the spots out was in sort of an artistic class. He was alone, and he did mystic things in the corner. The first man I saw like that wouldn't let you near him. He was like a magician. He didn't want you to know his tricks. Each one had his own private formula for getting spots out. There was a black man, who was a very good cleaner and spotter. He showed me how to do things. I remember, we had a pair of white flannel pants which they were wearing a great deal at that time, and we could never get them clean. He showed us how to bleach them. He took potassium permanganate and made a purple-colored solution, and dumped these pants in and they'd come out just purple. You'd swear they were ruined. Then you put them in a solution of sodium bisulfite and they would come out as beautiful white flannel pants. This was wonderful except they wore the pants out.

### Jill, dialysis technician, health clinic

I work in the kidney dialysis unit. I have to be there at 5:30 A.M. to start up the MAX, which is our central delivery system, and get all the machines ready for dialysis. Then the staff comes in and sets up the machines to do the dialyzing. We put out the stock for the rest of the day, and check the water for certain chemicals, and just mostly babysit the machines. Tuesday we get our supplies in, so I go in about nine. We unload the truck and carry the stuff upstairs and put it away. Whoever closes up the MAX and shuts down the system, flushes it with either bleach or something. We have a set routine, a checklist, where we're supposed to check off everything we do. Quality control. Make sure there aren't any mixups. If we did mess up and leave some chemical in the delivery system, it could affect the patients.

On Thursdays, after everybody leaves, I sterilize with formaldehyde. That's the biggie. Most of the staff doesn't come in contact with it. But I knew somebody who was sterilizing every week for over a year, and developed a cough and a wheezing that would last for a day or two at a time. I work with it at least once a week, and have a problem with burning eyes. I put four gallons of formaldehyde, 37 percent concentration, in the MAX supply tank. Then I pump it out through the system. There are 31 canisters in this room, they're all lined up, and I fill each of those with formaldehyde and turn on the recirculating pump. It has to run for two hours. Meanwhile the formaldehyde is just everywhere. There's no tops on the containers, so it gets in your eyes and respiratory system.

Right now, the artificial kidneys are disposable, We use them once and throw them away. But there's a new policy to start re-using them which means that all the kidneys will also have to be flushed out with formaldehyde between each use. That will involve more people using formaldehyde.

But I like the job. It's necessary to someone, because our patients can't live without it. Yeah, I like it. It's very responsible but it's not very difficult. One of the administrators a couple of years ago said they could train chimps to do our job. I don't think it's that easy. I was thinking of taking him a bunch of little chimps from the zoo dressed in lab coats and saying, "Here's the staff you wanted."

### Eric, sculptor, self-employed

I worked at a museum for four-and-a-half years doing sculpture; life-size nude figures, anatomical models, and other assorted exhibition work. I started in high school, doing brass welding, and then went to college and did a lot of wood-constructed pieces. In the middle of college, some 13 years ago, I went into work with plastilene, which I still use. Plastilene is an oil-base clay that is made out of clay flour, wax, oil, and a little bit of sulfur to preserve it. After I finish a piece of sculpture, I have a polyurethane rubber mold made by a mold maker. We cast polyester resin into that mold. The wax forms a hollow casting which is then sent to the foundry where they invest it with a special kind of ceramic shell material. Then they burn out the wax and pour the bronze into it. I'll work on the wax a little bit, and when I get the polyester resin (which is sometimes called bonded bronze, because you mix in bronze powder) I'll do the retouching. That means grinding, and adding little bits of resin, to patch the holes or seams. Almost every step of the way involves toxic chemicals.

When I talk to the mold makers, all I hear is chemicals. They used to use glue molds to make reproductions of sculptures. It's like an animal glue that's cooked in a double boiler and then you pour it on the sculpture and it sets up like a gel. You can peel it off and make two or three waxes for bronzes. Then the mold makers switched over to the bonded bronze because it looked like bronze but it's one-fourth the price. They switched from the gelatin to the rubber which was not smelly except that the polysulfide in it smells like rotten eggs, and how dangerous are rotten eggs? Then they eventually worked with plastics, and epoxy resins, and then polyurethane casting materials. Anyhow, they're working with chemicals all of the time. Polyurethane, silicon, and polysulfide. I think there's also lead in the hardener. They're there all day long and these molds are open. It's not like they finish a piece and that's the end of it. They brush into one, they brush into another, then they'll open that mold, brush, and make another cast.

Mold making was a real problem at the museum. The worst place in the world is where they make the reproductions of dinosaurs. They make latex molds from ammonia, latex and water. Since the dinosaurs have large surfaces and big molds, when the molders do the laminated polyester resin casting, they have a tremendous amount of surface that's giving off toxic fumes. I think it's the styrene which is the worst, and then the MEK peroxide which is a catalyst. They do have fans, but the place just reeks with that aroma. It's terrible, and a lot of people down there smoke and nobody wears gloves. They absorb the stuff into their clothes and then go home like that. They mix the cabosil, which is fumed silica, into the polyester, and the dust is all over the place. Then they grind the seams which puts the silica and the fiberglass cloth into dust form. I heard some awful stories about fiberglass dust working its way in through the skin.

### Debbie, hair stylist, beauty salon

I like to do permanents. Different perms have different things in them. Some are not as strong as others: the more expensive ones don't smell as bad, they don't take your breath away. They've come out with these new chemically treated perms, where you mix two chemicals together which makes the solution get hot. They smell terrible. Every time I do one, I get sick to my stomach. I haven't been able to find out any information about these at all.

You have to know what you're doing. You could burn somebody's head with a perm, or with a bleach or a color. Even though it's external, you could still hurt somebody. You could make their hair fall out if you're not careful.

The State puts up a safety list. Floors have to be swept and mopped daily, things like that. You have to wear sleeves, when you're working with a customer. You have to use clean towels; you have to have some kind of disinfectant for your combs and brushes and rollers. Everything has to be clean and sterilized. Most of these, though, are to protect the customer. The only safety regulation for the operator is that you have to wear shoes so that if you drop anything you won't hurt your feet. Dropping a pair of scissors, you could give yourself a nice cut. But as far as working with chemicals? Nothing. The rules are a bit out of date.

We use precautions, but they're also geared to protecting the customer. When you put the solution of a perm on a head, you put cotton around so it won't drip in their face, or get in their eyes. I have to tell the new girls all the time to take the cotton off after it's soaked with solution because if you leave it there, it's going to burn the skin. But I also insist that they wear gloves when they do a color or a bleach. If they say, "Oh, I don't want to," tough, they have to. We also have a very good ventilating system that's on all the time. So even though it's kind of confined, it works out okay. But no matter where you do a perm, even right underneath a fan, you can smell it. It smells terrible. That's just the way it is.

### Bill, fire fighter

Fire-fighting is a dirty job. We work 10-hour days and 14-hour nights. The average day on the fire department begins with a roll call at 8:00 A.M. After that you check your personal safety equipment. There's a whole routine you go through everyday that's sometimes interrupted by fires. We have housework to do, because it's a 24-hour habitat. It's a home away from home, like a big fraternity. Housework and preparing the noonday meal become something of a community activity.

The station I work in is like being on submarine duty. There's no windows. It was designed by the Marquis de Sade. It's easier to call home to find out what the weather is than to look through the opaque windows and figure out what's going on. Most stations are like a house. You wouldn't consider them more hazardous than an average home except for the fire trucks parked in the living room.

But our main workplace is a hostile environment. We try to find out what's in a building before going in. If it's an apartment house, we've got a good idea. We inspect factories so we know pretty much what they're carrying. We make a point of hitting every commercial building in the city every year. But you never know.

One of the worst situations was the state office building fire. It began like a typical fire, an alarm going off in the building. We sent one company to investigate since it's usually just a malfunction in the alarm system. When the apparatus left the building, we were toned for what's known as a "telephone alarm," that's a long tone and means you're going somewhere. The call came in that we had a fire in the control room in the basement level of the state office building. You know it's a fire, but you don't know until you get there what you're going to have. Your heart's going tunk, tunk, tunk. . . . We pulled up and there was smoke coming out—you could smell it. This fire was unique for me in lots of ways. We thought there were explosions. Now we know that what really was rocking the building with loud noises was the generator for auxiliary power. There was so much smoke in there that the generator couldn't get enough air for complete combustion, so it was violently backfiring. Also there was a lot of electrical arcing that would physically blow open the doors to the control room. The building was shaking. Standing outside the door was, you know, macho men don't like to use sissy words, but it was SCARY with those doors blowing open. We had to wait for the gentleman from the electric company to come because there was no way we were going in with this thing jumping all over the place. So we set up a staging area, brought in the equipment, and had it all set up for an initial attack while waiting for the guy from the electric company. He shows up and says "Geez, you've got PCBs burning in there. I don't want to go in there. I just bought these boots, I might be buried in them."

I said, "Whoa! What are you talking about?" He told us that the stuff causes cancer. Here were all these guys standing around in smoke, covered with this oily soot. It was all over the place, foul and rancid and acrid. We didn't even have our masks on because we still hadn't attacked the fire in the control room. It was a terrible smelling smoke, but to the eye it was clear. We just weren't prepared.

The fire itself was nothing as fires go. Once you shut the power down, all you had were some class A combustibles. But then we realized that this wasn't just cooking oil on the stove. I asked the electric company what they do when his people got into this. He said, "Well, we take off all our clothes, and seal them into metal containers, and they ship them off someplace to a landfill. Then everybody takes a thorough shower and they check our blood at the hospital." That's the procedure we followed. We had a full line there at the time: three engines, two trucks, and a squad, and we immediately got everybody out except the squad. We left all the hose lines and everything else right there. As soon as we got back to the station, we stripped right down, I mean everything. We didn't have metal containers, so we used plastic garbage bags and sealed those. We took showers and went off to the hospital for blood tests. That's when we ran into some problems. We got a runaround and still don't know what's going on. They're dragging their feet because they don't know, so they're watching what's happening to us. They sent us to the hospital for blood tests, and sent us back again. It almost seems that they're using us as a case study.

They found not only the PCBs, but also dioxins. Apparently this pyranol—that's the generic name of the substance—was heated by the high voltage blaze until it broke down into dioxin. Fortunately outside of what sticks out around your face, we had complete coverage with gloves and boots, all the way up. But we were there; it was like 1500 to 2000 degrees in that room. So this stuff was not in the inert state like it is for the people who come in there to work. It was flying through the air, and it did penetrate, because we had soot and stuff right up our pants and our sleeves. Some people had chemical burns.

The state health commissioner said, "We've got to shut this building down right now, seal it off, make sure none of this stuff leaks out." And here we're saying, "Well, what's it going to do to us," and not getting any answers. At first there was a lot of fear. We had no idea, I mean, we were crawling on our hands and knees in this stuff. We didn't know that the transformers were filled with pyranol. They only put a little warning sticker on the side of the transformers. They were in a corner and it was burning and everything was black and full of smoke. I'm sure that the problem would have been minimized if we had known what we were up against, if we knew what was in there, if we knew what was happening.

# PART 2

# Problems on the Job

# 2    Illnesses and Complaints

Over a hundred years ago Lewis Carroll acknowledged the cerebral effects of mercury poisoning in the character of the Mad Hatter. The health effects of lead have been recognized for hundreds of years. Carbon soot was associated with scrotal cancer in chimney sweeps in the mid-eighteenth century. Coal and silica dust have long been known to cause respiratory disease.

Prior to World War I the dusts from the extraction of raw materials and the fumes from the processing and use of basic chemical substances were common health hazards. Then, in the 1950s and 1960s, changes in technology brought new and unfamiliar problems. As the chemical industry matured, production shifted to synthetics and plastics, and even older products such as steel and glass were reconstituted through new chemical processes. While preventative measures have reduced exposure to some familiar substances, such as lead, rapid technological changes in chemical products and processes have created new hazards.

Following policy changes during the 1970s, which created greater awareness of occupational risks, those working with hazardous materials are increasingly inclined to associate their health problems with their jobs. However, it is often extremely difficult to make definitive causal connections between illness and work. Relatively few chemicals have been adequately tested to identify exactly what problems are due to what levels of exposure. Data on the effects of long-term, low-dose exposures are especially elusive. Intervening effects such as individual variation in response to exposure, the synergistic effects of multiple exposures, and differences in life-style, diet, and smoking habits can obscure the relation between work and health. Cancer, genetic disease, sterility, and other health problems have multiple interdependent causes. Work can be a direct cause of an illness or merely a contributing or aggravating factor.

Understanding of occupational disease is further obstructed by the lack of sufficient information. Data from animal experiments or human experiences are scarce. In some cases, production levels have increased so recently that long-term effects cannot yet be known. But even in the case of older chemicals, large gaps in the employment and medical histories of workers, the failure to report health problems, and the weakness of diagnostic techniques have precluded the systematic collection of information. Epidemiological studies (identifying patterns of disease to determine causation) often are frustrated by inadequate historical data. Long-term studies are costly and rare.

In this chapter, workers talk about what they believe to be the health effects of working with chemicals and describe why they draw associations between their health and their work. We have listed and described the illnesses they report in Appendix 2A.

**Symptoms**

Few of the workers we talked to had personally experienced traumatic problems, for they were relatively young.[1] Most of their complaints are about persistent skin irritations or respiratory ailments. People rolled up their sleeves and showed us rashes and chemical burns; some of them wheezed and coughed during the interview. They described routine symptoms such as allergies, numbness, disorientation, and respiratory problems that they called a "chemical cold." Because of the latency period for chronic disease, younger workers have fewer chronic complaints than older workers; most of the cancer cases we heard about involved older employees. Consistent with the findings of the Quality of Employment surveys, unionized workers were more aware of health hazards than nonunionized workers and more likely to feel that their injuries were related to the job.

Some workers discussed problems with us that they had not reported to their employers. They felt they had inadequate proof and expected that their complaints would be disputed or dismissed. Some feared that reporting problems would lead to retribution. A number of workers dismissed their ailments as personal problems. Maintenance workers, railroad workers, artists, and others who have little interaction with their colleagues often be-

1. The age structure of our respondents in the chemical and related industries approximates that in those industries nationwide as reported by the U.S. Bureau of Labor Statistics.

lieve that their problems are unique. The embarrassment and dread of certain problems such as sterility, cancer, or nervous disorders make some people reluctant to share experiences. Illness is often a lonely affair.

---

### Sandy, rigger, chemical plant

When I was exposed to mercury, I suffered all the symptoms of mercury poisoning: the tremors, the blurring of vision, the dizziness. I couldn't sleep at night; I'd go without sleep for a couple of days. I would dribble at the mouth like a baby, and wet my bed. I was irritable to the point where I became violent at times.

They wanted to give me something that would reduce the mercury level in my system but I didn't want to expose myself to another substance when I'd already been poisoned by one. I just didn't believe in that. I thought that if my system couldn't naturally rid itself of this foreign substance, then I would be in trouble. Because of my irritability, dehumanization, and tremendous mental stress, I also placed a burden on my wife. Then my daughter was conceived when I had mercury poisoning, and six weeks after her birth she suffered from seizures that were unexplained. I believe that she had mercury poisoning too. So it was not only me who had to pay the price for the money that I earned, but my family suffered in order to enjoy the benefits of a good paying job. The company told me they were sorry, that I could move to another job, away from the mercury. The Big Brother concept—don't worry, they'd take care of things and I'd be all right.

### Sheila, laboratory technician, research institute

My work with fungicides has caused some destruction in my lungs so that I have trouble breathing every time I go out into the cold or whenever I'm exposed to strong odors or too much exercise. I had to go through a really conscientious program of redeveloping my lung capacity, and that's not easy. I mean, that's something that's never going to leave me. That's like going somewhere and all of a sudden developing asthma. It's a part of my life now, that didn't used to be a part of my life.

### Joe, laboratory assistant, chemical plant

In the years I've been here I've seen people get hit with just about everything from cyanide to phosgene. I once spent 30 days in a hospital when a phosgene tank blew up on me. We had the phosgene in 2000-pound cylinders and ran it in through hoses. There was a leak in the cylinder one day so I notified the supervisor who said I should go out and tighten it up. But when I twisted the wrench, the whole fitting fell off and the phosgene hit me. I got a pretty good dose of it. I went into convulsions and couldn't breathe. They had me in an oxygen tent. Then, for months after I got out of the hospital, every once in a while I wouldn't be able able to breathe. That's happened two or three different times in the last eight or 10 years. The day after the accident I was given a written warning from the company because I didn't have the proper safety equipment when I went over to the phosgene tank. You know, they covered their ass.

I've also gone back into the hospital with massive headaches, but nobody can ever find out what they are. Every so often I'll wake up in the middle of the night and feel like something has just kicked me in the head. A couple of neurologists told me that it may be linked with the phosgene, but it's something that you could never prove. There's no sense fighting it.

### Eve, sorter, manufacturing plant

We have one girl, she's about 41 or 42, working at the company who has some kind of sickness that ruins her muscles and nerves. She has numbness; she has a hard time walking. Her legs are so weak that she has to use a cane. I believe it's because she's worked with hot triad (trichloroethylene). That girl is one mess.

### Eric, sculptor, self-employed

There are a lot of stories about artists working with polyester resin. I have a friend who became so ill from it that his doctor wouldn't allow him to go into the city for more than three hours at a time. He can never mix resin again. G.—he does realistic work—I heard he had cancer, then I heard he didn't. Now I hear that he has it again. It's frightening. There was a mold maker who was doing laminated polyester castings, giant ones, in a room without a fan in the window. He was doing this for six months, and he started getting sick. They didn't know what was wrong with him, and he got worse and worse. Then he finally started getting a little bit better, but I think there might have been some permanent nerve damage. Now if I read on a can that polyester resin deteriorates the nerves, I feel the danger is very real.

### Dorothy, deckhand

I've seen some pretty batty people on boats, old guys with twitches and all kinds of weird stuff. I mean, who knows what it's from? But it makes you wonder about heavy metal exposure and things like that.

### Debbie, hair stylist, beauty salon

Six years ago, I came down with a lung problem that they thought was TB. They kept me in the hospital for two weeks and put me in isolation. They found out that it wasn't TB, so I went to see a lung specialist. He sent a scraping from my lung to a special lab in Virginia that deals with the kinds of diseases that hairdressers get. They call it "beauticians' lung." It was a fibroid tumor from using hair bleach. I used to do a lot of coloring work, which I don't do any more. There's just no way of controlling it. It happens to one in maybe a couple of hundred, very small odds, but I also smoke, which most hairdressers do as well.

### Fred, chemical operator, chemical plant

I had cancer of the mouth. I don't know if it was from work. The doctor asked me if I ever smoked a pipe. I did, but she says that's not a likely cause because I quit years before. I told her I work in a chemical plant and that we use a lot of chemicals there that make your mouth dry. We're all the time having chapped lips. She says that could well be the problem, but I'd have a hell of a time proving it. They operated and I was cut from one corner to the other. My lower lip's numb now.

I've noticed quite a few of the guys have chapped lips in the winter, and skin problems: a redness, soreness, rawness of their face and neck. They blame it on this HCl (hydrogen chloride) and the cold which tends to dry out your skin. You see, our plant isn't heated. We have a control room that has heat because there's instruments, but, aside from that, the plant area is right out in the open. This winter was real tough at 10 below sometimes.

People also have problems with their feet. It's a fungus. A specialist took a scraping from my foot. I've had this about 12 or 13 years. They're like little water blisters. They itch like crazy and I break them with a needle and put on a little iodine. Four or five other guys have it too. It comes from this stuff on the floor. You get it onto your shoes and it eats through them. One time we were loading cyanacetic into this stainless steel tank, and afterwards one of the guys grabbed a hose and washed all the cyanacetic with alcohol in this big funnel. Somehow I got some of it on my shoes. It was potent stuff, and I got a hell of a burn on my foot. Didn't even realize it until later. That's a problem here. You get exposed to some hazard and you don't even realize it.

### Pat, graphic artist, community agency

Often, I don't know, I just feel kind of in a bad mood and don't care, or snap at the people who come in to pick up their printed stuff. It isn't that it's not a good job, not satisfying artistically. It's a job that's in a lot of demand. I think that it's the chemicals. You know, it isn't that you would call it an illness, like a rash or anything as serious as lung disease, but just a constant vague headache. Plus the stuff just leaves a film on your body. At the end of the day you can take a tissue across your skin and there'll be a light coloration from the ink you've been using. The inks seem to vaporize, so you are breathing it. I wouldn't bring my baby down here. Other people bring their kids down if they work late hours or if they're working at night. But I wouldn't want him down here, with that stuff.

### Mike, photo lab processor, blueprint company

I've been exceptionally healthy, knock on wood. I'm just that type, I don't know what it is. If anything, I'll get a cold once a year. So I've been darn lucky. But ever since I've worked here, I've been sneezing, coughing, and you know, my mind is screwed up. I get fuzzy, I talk sometimes and I'm all clouded up. Half the time I just can't think straight. Now I don't know if it's because of aggravation or chemicals. I have a little chemical book and it mentions something like loss of memory. But there's another word they use to describe it, like absent-mindedness. Something like, you just walk around in a fog all day. And I do. I guess I've been here for roughly 2¾ years, and since I've worked here I've always had this feeling of being kind of blaah. You walk around, you just kinda feel like a weirdo, on some kind of drug.

### Sally, services technician, hospital

Two friends of mine who work in a respiratory therapy department use Cidex to sterilize their instruments. It really bothers their respiratory tract a lot. It gives them trouble breathing. They recently changed the product, giving it a minty smell so it wouldn't be quite so obnoxious. But more people have complained since it has had this mint odor. It sticks in their throats a lot more and makes the eyes burn.

### Nick, chemical operator, chemical plant

I haven't been real sick, but we all just get what we call a "chemical cold," constant hacking, inflammation of your arms and neck. It makes you feel like you've got a cold. They have had chemicals here where the fumes get loose and your eyes would close up on you. You feel like you have sand in them, and you can't look at a bright light. We've had chemicals like that in the past. We no longer make them. Thank God for that.

### Les, furniture restorer, self-employed

One of the tests we do when we're getting the oil out of the table top uses a thing called a "spirit rubber." You put alcohol on the rubber and you test it by putting it up to your mouth; there's a certain feel that you get with your lips. That's the traditional way of doing it. It has been that way for years, since French polishing was introduced in the early nineteenth century. You basically have a bit of a lick at the rubber and see how it feels. It's soaked with methylated spirits. I know a lot of drink-crazy polishers who are just drunk all day. I'd be interested to know if the constant exposure to fumes and the testing of the oil gives you a desire for more than the average.

### Kitty, industrial painter, university

I got a rash so bad all up and down my arms. It itched, it was painful, and I couldn't get rid of it. I finally figured that it must be damned solvent that I wash with three times a day. I approached the foreman about getting hand cleaner and he said, "Aah, you don't need that." He finally did come across with some, because I complained so many times. In the meantime I had gotten a can of hand cleaner for home use, so lots of days I'd just go and clean up at home. I couldn't get rid of the rash. It just wouldn't go away. There's one guy on the job who has awful trouble with the skin on his hands. It just comes off in sheets. I've never seen anything like it. I says, "How come you're a painter, I mean, why don't you go do something where you don't have all the solvents on your hands all day." He says, "I like to paint, it's the only thing I know how to do." His whole family paints, everyone. Boy, I wouldn't do it if my hands fell off like that.

It's funny but painters, nearly all of us smoke and drink. I'm on a crew now with 10 people and eight of us smoke like chimneys. You can't help but think that the fumes have something to do with it. When you're using a pigment in shellac which has an alcohol base, it has a very overpowering smell. In an enclosed space it can knock you out. I've spent time in small kitchenettes and bathrooms when I knew I should've gotten the hell out sooner than I did, but I wanted to finish the job so I didn't have to go back in. It's funny, but the cigarette smoke will kill the smell. It'll kill the taste. It's probably why so many painters smoke. Painters also drink a great deal. You know, the joke around the union hall is that when you need a painter, he's either a drunk, on his way to becoming one, or he's reformed. Chemicals dry you out, the dust, the taste, you have to get the taste out of your mouth.

I once spent two days cleaning out brushes, a couple of dozen of them, all hard and gooky. I cleaned them with lacquer thinner. I drew it out of the big 55-gallon drum in the back, and for two straight days I squatted over a bucket of lacquer thinner cleaning out brushes. By the end, I was pretty sick. Real dizzy, headaches, and I didn't dare drive home. So I drove as far as the bar downtown and had four or five drinks, until I felt as though I could drive. Now it's weird that you would have to sit and drink beer in order to be able to drive, but I had to get it out of me. I don't know if the volume of fluid and getting the chemical out through the urine has anything to do with it or not, but I sure did feel better.

### Ken, electrician, chemical plant

I worked with PGCH for three or four years. At first I never had any problems except for a runny nose once in a while. I took that as part of the job. But then all of a sudden an allergy developed where I couldn't breathe. If I was exposed at all to it, I became an emergency room case with a severe asthma attack. The doctor recognized it right away. I'm so allergic you could use me as a barometer. If someone were to sprinkle a little PGCH in this room, you wouldn't even smell it and I'd get sick. One time I took a short nap in the locker room after I ate lunch. I couldn't breathe when I woke up. I sat there for a while, then had someone get the nurse. The nurse and the safety director gave me oxygen. I told them there was PGCH in the air, but they said there was none in the plant. But a friend found out that only about 100 feet away from the locker room they were drying it and emptying it into the air by spraying it out with a fan. I was off for two weeks with Benedryl pills and antihistamines.

Now, if my wife does windows with ammonia, I can't stay in the house. It wipes me out with a headache and sinus attack. I know that household ammonia is a very caustic agent, but I don't think I should react like that. The doctor said I'll always be more sensitive now to a lot of things.

---

### Associations

People who experience illness want to know its cause. They search for causes for practical reasons—to get compensation or to protect their future health. They also feel a personal need to know, a need to define the origins of their problems, to identify a source. Establishing cause is a comment on the social system, a way to define responsibility and to assert control. However, technical uncertainties create problems for workers who want to document their experiences. Often unable to gain access to the data necessary to evaluate hazards, they find it difficult to isolate cause and effect even when they suspect the workplace is the source of their problems. And they also find it difficult to convince others of the validity of complaints that are not fully supported by the scientific literature. This restricts their ability to make decisions about continuing work, to substantiate requests for better working conditions, and to document their compensation claims.

Despite such difficulties, workers do make associations between health and work by observing certain patterns. They may suffer specific symptoms only during certain periods of a week: "I'm fine on Monday, but by Friday I'm ill." They may notice that symptoms are associated with certain locations or substances: "I'm sick every Thursday when they pour chemicals." Or they may discover through reading or in conversation at the union hall or at a social occasion that their problems are not unique.

A Labor Department study, analyzing the reliability of workers' self-reporting of hazards to ascertain occupational causation, finds such reporting to be consistent with scientific sources of information (except in underestimating the effects of chronic disease). And in fact, most occupational health hazards have been uncovered by the people exposed to them. Yet there remains considerable distance between the personal experiences of workers and their ability to garner evidence that meets scientific standards of proof.

---

### Vivian, laboratory technician, research institute

I had a lot of skin rashes from working with chlorox, and sometimes I'd experience eye problems from exposure to some of the dusts. In fact my eyes were generally very tired and irritated because these chemicals were being blown in my face all day. After a short vacation, all the patches on my skin would clear up. But when I'd come back to work and get my hands in whatever I was working with, it would act up all over again. A lot of people had that happen. You'd get headaches or nausea during the day, then if you were out of the lab for a period of time, even just a weekend, you'd feel great. Then by Friday you'd feel like hell. That's how we were able to tie it to the job.

### Ben, repairman, chemical plant

I was always itching. I believe that it was from work. It didn't itch when I wasn't there, but only when I went to that section. When I went there to work I started itching all the time. Not only me, but everybody else's skin itched too. I felt that I shouldn't have to take that, but then, well, Jesus, it's a chemical plant and the lime's there and part of the process and no one is going to get rid of the lime. You've got to keep that plant running. So what can you do about it except make sure that everyday you put your cream on. That usually eliminates the problem. But still you're putting cream on because of the situation at work. I didn't take safety and health seriously until I started realizing my own problems. And then when I saw nothing being done about it, I started getting involved. Actually it was safety that I got involved in first, not health; you know, the steam blowing out. Health wasn't really an issue at that time. We all thought, so what if you start scratching when you start working on a job.

### Mike, photo lab processor, blueprint company

I got this book and looked up the chemicals I knew we used at the lab. It mentions things like dizziness or loss of memory. You're just kind of out in space someplace. It also mentions rapid breathing, shortness of breath. I'm thinking to myself while I'm reading this, "Oh, my goodness, this is just the stuff I'm experiencing!" Well, I'll give them a fair chance, you know, maybe it's just me. But to read this after you've experienced all this for two years . . . I used to think, well, maybe it's just me, maybe I'm getting older, maybe I'm just in a bad mood. Well, after reading this, it's very hard to shake that feeling that something's up.

### Kitty, industrial painter, university

I was reading a book and ran across something on the symptoms of chronic lead poisoning. What it will do to you is restrict the blood vessels to the kidneys so that the kidneys absorb more salt to raise their blood pressure. At the same time they raise the blood pressure over the whole body. I read that and thought to myself, "Jesus, you know, S. had high blood pressure, and he's been painting since 1950 when they used lead paint." Then I thought, "Gee, P.'s got high blood pressure too and he's been painting since the forties." The cases just started multiplying. The more I thought about it, the more painters I knew with high blood pressure. There's got to be some relation. Out of 15 painters in our crew maybe 10 of them are old enough to have spent a lot of time with lead paints, and out of that 10, six or seven have high blood pressure and nobody ever connected it. But you can't prove it. How are you gonna prove that it's a job-related disease? It's impossible. You go to the doctor and he'll tell you, 'Well, you're 10 pounds overweight and smoke too much, that's why you've got high blood pressure." They don't connect it with the lead.

### Mary, housewife, wife of railroad conductor

The feeling I have is that "Don" had dioxin in his system when our baby was conceived. You breathe that stuff, and it goes in through your pores. There's a genetic tie-in. I don't really want to pin her cancer on any one thing, yet there has to be one thing that caused it. I can't believe that there are that many different things. She didn't inherit it from somebody else, though they used that excuse when she died. When she was sick there was never any discussion about where it came from. It's not a thing that anybody would usually think about it, unless you work in an asbestos factory or with radiation. We had never thought about dioxin. A month and a half after she died, an article appeared in the local newspaper about the effects of dioxin and its use in weed killers on the railroad. We read the article and then that night we went back and read the crew books. Don has all this documentation because he had to write down where he worked everyday. The dates coincide perfectly. Horrifyingly perfectly. The paper wrote it up in big block letters: THE RAIL-ROAD AND DIOXIN. There it was, soft-cell sarcoma, the whole thing: it was written right there. Word for word, her diagnosis. That's when Don called the company.

### Sheila, laboratory technician, research institute

We had contracts from major chemical corporations to do pesticide screening. The corporations would develop the pesticides but wouldn't test them on the bugs, so, we raised the bugs, and screened them. We would do maybe 10 or 12 pesticides a week to screen for their efficacy. They were obviously affecting everyone in the building. We would be working on the first floor with stuff that smelled real strong and you could smell it on the fourth floor.

I was sick everyday that I worked in that place. For two years I was sick. But I mean, sick, like subtle sicknesses. Like feelings of disorientation and not being able to focus my eyes properly. And this feeling sometimes that I would have to steady myself to walk. That began within three weeks of starting work in the lab.

They know that I wheeze and that, yeah, there's some problems, but I can't prove that those problems didn't exist before. How could anyone know that? I mean, it looks like asthma. The only difference is that asthma is reversible with medication and this isn't. But I could have been exposed elsewhere. . . . Who knows that at night I didn't go home and sniff glue, right? It's possible. But I know, because we were mainly getting sick every Thursday, and Thursday was the day they poured the chemicals.

### Peter, railroad signal inspector

My son was born with one foot out and one foot in. He had to get a bar. My daughter was born with dislocated hips and both feet pointed in. She was put in a body cast for three months. She was in traction for a week when she was three months old. For a three-month-old child, it wasn't a good experience. And she has more problems than just her hips. Everyone says that she's extremely intelligent but she just can't get it together. She's hyperactive and she's on almost 100 percent natural diet. No additives, no preservatives, no coloring. We took her to a special school, and they asked me if I had been in Vietnam. I asked them why they asked me that. They wanted to know if I was exposed to anything like Agent Orange because they're finding that a lot of children in this school have a father who was exposed to something. I told them that the railroad uses a weed killer. At the time I didn't know what it was—that white stuff that you see along the tracks. I still thought it must be my own fault. But then at a union meeting I found that I wasn't the only one. When you're by yourself you just don't really put it together. I wasn't sure of what they were using. Until my friend talked about his problem with his kids at the meeting, I didn't know that the spray is the same chemical as in Agent Orange. It made everything click. I just put it all together. Before I didn't think that one had anything to do with the other. Maybe I should have, but you don't think along those lines.

### Bess, diffusion analyst, manufacturing plant

I went to the doctor because my eyes were very inflamed. He said it wasn't my system, it was an outside thing that had infected me, but he couldn't tell what it was. I had quite a problem with it. I brought it to the foreman's attention. I brought it to the engineer's attention. Finally I ferreted out the problem myself. The vent wasn't correct. The fumes were going back into my eyes. I didn't recognize it when it happened because you can't even smell it. That's the bad part. I found out because when I was ill, M., who works with me, told me that she had the same problem with her eyes. It was at that point that I knew it had to be the vent. It wasn't logical that the two of us should have the same problem.

I'm very suspicious. We've had a number of people with cancer and we brought it up last year to the front office. They said they researched it and decided that there was no correlation between the people who had cancer and the work they did. We've asked them two or three times for that research report and so far we haven't been able to get it. So whether it was really done, I don't know. I think they took it too lightly, so we started writing down everytime somebody had been sick with cancer. We talked with some of the ladies who had been there a good many years and made a list. We have more than 60 people on our list since 1961, and I'm sure the list isn't 100 percent because we just did it informally. It seems to me that's a lot.

### *James, computer assembler, manufacturing plant*

We hear of cases where people in the same department have the same type of health problems. There might be people working in one building with one chemical all getting esophagus problems. But then we can't document it because they just shift these people to different areas so it looks like there's no problem. The workforce never knows what's going on. In one section where people worked with cindered metal they were having a lot of problems. One guy got emphysema. Another guy worked in an area that used epoxy fibers in between the printed circuit boards. He died from lung problems. From what I understand, it's a lot like asbestos fibers, and he had spots on his lungs, even though he didn't smoke. His wife also works there, and they gave her a good job and some money so she doesn't want to talk about it. Our biggest problem is getting people to talk.

### *Nora, graphic artist, blueprint company*

I was just falling asleep one night and got thinking, wow, there were so many people who have died where I work. Two people just died and two people now have cancer. The year I came there, a guy died of leukemia. Two years later, a kid who worked in the store died. And now, a girl in the offset department has cancer and a guy in blueprint has cancer. She has cancer of the esophagus and his is in his lungs. J. had it on his forehead. Now a third person has been having this cyst drained and has to have a biopsy. I thought, "Wow, that's a pretty high incidence." I wanted to do some statistics on it, to see how it compares with the national level. I've called the Cancer Society—a friend of mine's the director of education—to get some stats.

### *Ken, electrician, chemical plant*

I could name 10 people here who have died of cancer. My supervisor died but I really don't know if it was caused by the job. He worked for many years with melanic acid and toluol. An electrician who died of leukemia swam every day in benzol (benzene). The men in his shop used to rinse their hands in it. It got in their shoes, so they absorbed it everyday. They walked in puddles of it.

### *Steve, railroad trackman*

I heard that Agent Orange or some such thing was used to defoliate railroad tracks. That rang a bell because I knew something about Agent Orange in relation to the Vietnam veterans. It also occurred to me that we were getting a lot of medical claims submitted by union members for their newborns, and that didn't seem quite right. Then I made the connection. So we started going through our health and welfare records for the railroad membership. We didn't have very complete records but they were relatively complete for 1978, and we found that roughly a third of the newborns had some kind of birth defect, most of them being some kind of foot or leg deformity.

### Jocelyn, secretary, museum

I started working at the museum about a year ago and immediately got various ailments that I never had before. I had a whole lot of trouble with my eyes, and I was really dizzy, like my feet would hit the ground and it wouldn't be where I expected it to be. I thought it was stress because I hadn't worked full time for a few years I just explained to myself that I wasn't cut out for full-time work and was having a hard time adjusting to it. But then in the middle of the winter I got a rash that covered half my arm. The dermatologist didn't know what it was, but said it looked like a reaction to something. Then two other people got the same kind of rash and we all worked on the same floor. That made me think that there was something in the air.

A few months later a man came to reaccredit the museum. He found that a thick oily film was being deposited from the air on the art. He advised us to send a specimen to the lab because it could damage the varnish. The report that came back from the lab said that it was diethylaminoethanol, a chemical that's in the humidifiers. It said that the highest amount you should absorb in the air is 15 ppm and we had more than that. It said this chemical was an eye and skin irritant and that people using it were supposed to wear gas masks and gloves. I started asking everyone if they had any symptoms and found an incredible number of people on the staff with dizziness and eye problems. The people who wear contacts had to clean them every hour because the same film that formed on the paintings formed on their lenses. Other people who thought they were getting hay fever now began to think that it was this chemical. We were talking one day at coffee and came up with 17 out of 42 people who work here who had mysterious rashes.

# 3    Anxieties and Fears

"Chemophobia" is a pejorative term often used to describe the growing public concern about toxic chemicals in the environment and the workplace. A "phobia" is an irrational and illogical fear. It is to be distinguished from anxiety, which has an explicable and rational base. Anxiety stems from an anticipation of problems, especially those which cannot be avoided or controlled. It is a serious and pervasive problem for the people we interviewed. Even those too young to have had any personal experience with degenerative disease are bewildered and worried about the effects of their work on their health. Conflicting pressures compound their concerns as their need to maintain a job confronts their fear about the effects of exposure at work. Far from a phobia or irrational fear, their anxieties about working in a hazardous environment reflect their limited ability to control the conditions of their work or to find alternative jobs.

People worry about certain diseases more than others; in particular, cancer and genetic problems evoke a special dread, though their incidence is less immediately evident than many other ailments. In a penetrating essay, Susan Sontag suggests how certain diseases are prone to symbolic interpretation. While illness itself is a biological reality, diseases that are poorly understood are often seen as more than just diseases. They become a metaphor for social injustice or repression and, for the workers we interviewed, a symbol for broader concerns about the nature of their work.

Corporate policy has virtually ignored the problem of anxiety, dismissing it as unfounded or phobic. Though some discussion has recently come to focus on the effect of stress, this has been defined in terms of such factors as noise, repetitive work, and strained interpersonal relations. Government policy has also ignored the problem as irrelevant to regulatory proceedings. However, in a 1982 case involving environmental rather than occupational health, citizen groups demanded that the NRC decision to reopen the Three Mile Island nuclear power plant consider the psychological effect of anxiety and fear about risk. They argued that effects on psychological health are effects on the health of human beings. Although the Supreme Court ruled that the NRC could limit its considerations to physical health effects, similar challenges are conceivable in the context of growing concern about chemical risks.

This chapter conveys the terms in which people express their growing anxiety about their exposure to chemicals at work and its effects on their health.

### "It Scares the Shit Out of Me"

Many workers who once disregarded odors as obnoxious are now inclined to regard them as dangerous fumes. They express their worries and persistent fears: "It's always on the back of my mind." Sometimes anxieties extend to products that are not especially toxic, but this often reflects lack of information. Ken, for example, confuses dioxin and dioxane, suggesting the relationship between knowledge and anxiety. Like the people living near Three Mile Island or Love Canal, some workers feel that the psychological and emotional stress of being exposed to chemicals is a problem in itself: "You're dealing with a person's mind." Emotional stress is exacerbated by the lack of power to prevent further exposure, and the need to confront agonizing choices between personal or family health and earning a living.

*James, computer assembler, manufacturing plant*

To tell you the truth, all this stuff scares the shit out of me. For the longest time I worked with the stuff without knowing its effects. I have to admit, we were pretty sloppy about what we were doing. Nobody warned us. We used to clean off the glass that we put over the copper panels with methyl chloroform. We had a tub and a rag, and we just pushed a plunger down, soaked the rag and washed it down. The fumes would hit you in the face. We would do this for eight to ten hours a day. We knew it smelled, we knew it was a chemical, but we didn't think much about it. It wasn't until I started reading more that I realized, "Geez, this causes cancer problems, liver problems, kidney problems, and everything else.'' The fact that I've worked here for nine years with all kinds of chemicals makes me nervous. Shouldn't I have been told a lot more by the company so I could look after myself? The OSHAct protects us, but the company gets away with what it can. If we do things because we don't know, we can blame ourselves, but we can also blame the company, because it's not taking a real interest in informing us. Somehow management's afraid that if we know too much we might refuse to work.

I think what bothers me the most is the uncertainty of whether I'll make it to the end, or whether some stupid methyl chloroform spill is going to get me. I don't know. Whenever I'm ill, I wonder, have I been around one chemical too many? That stuff has gotten in me and isn't going to leave. From what I understand, it accumulates. Plus you have three or four different types of chemicals and nobody really understands the interaction between them. Paranoia on future health hazards is a big one. My wife always tells me I'm a pessimist, but when it comes to this I'm real pessimistic.

### Laura, filter cleaner, pharmaceutical plant

I haven't had any really horrifying health problems from the job, but there are things that I question. Sometimes we have to wear cartridge respirators because of all the nice stuff that's coming out of the exhaust. The first time I did that, I was really frightened. I stood on the roof inside the door and was shaking, just waiting for some wacko scientist to turn his hood on and kill me.

We had a job once where my boss told us to change these bag-out filters. We had to put on a bunny suit and a respirator. It was some really sickening drug in the filters, and I almost croaked. I thought, "There's no way I'm gonna do this! I don't wanna die." It was just like he was telling me to be so careful that I was terrified to even attempt to do anything. It turned out we couldn't do the job anyway.

### Greg, air-conditioning repairman, university

The more I go into these labs, the more I worry. It's always on the back of your mind. Whether you think about it or say something about it, it's always there. With all of the bottles and all the chemicals, you just don't have any peace of mind that these people are doing their job correctly and if they have an accident that they clean it up. Being involved with heating and ventilating, I know what a lot of these exhaust fans are putting out. I know that some rooms that are supposed to have positive air filters have negative. The cancer research rooms worry me, but for some reason genetic research worries me more.

When I came here two years ago, the supervisor told me a story about a professor doing research on cancer of the liver in turkeys. He told everyone there was no way that humans could contract this type of cancer. Six months later he died of liver cancer, and shortly after that his assistant died of liver cancer too.

### Ted, welder, chemical plant

When I came here to work I was young and did what I was told. The guy that broke me in said, "I've been a welder here for 25 years; look at me, I'm fine." But now I'm afraid of fumes. Everywhere you go, there's fumes: ketone fumes, toluol fumes, sulfuric acid, metal fumes. When I weld a tank that's had chemicals in it for years, even though the tank has been steam cleaned, you're actually burning off the metal, and you liberate fumes that have been absorbed into it. I worry a lot. If you just read the labels on the cans for welding runs, they have things that I'm not familiar with but have been breathing for a long time. Everyday, you smell something different. There's always some kind of stench in the air and you're never quite sure what it is or what the concentration of it is or what it's doing to you.

### Ellen, laboratory technician, chemical plant

I've been here 13 years working in chemicals. On an everyday basis, it doesn't really bother me. But when you sit down and think about what problems chemicals can cause, you just wonder, "God, is this really affecting me? In twenty years is something going to show up?" We have one chemical that women are not allowed to use anymore. They think it causes birth defects. We were told by my supervisor not to use it. So if we are analyzing the particular product, we have to ask one of the men to pour this chemical, and when you handle it with the solution before you analyze it, you wear gloves to transfer it from one piece of equipment to another. This is probably not enough protection, but seeing that I had worked with it for maybe five or six years before I knew, the damage that could be done was probably already done . . . it really makes me scared, you know, and sometimes I think, "God, why do I stay here?"

### Ken, electrician, chemical plant

When I first came to this company 16 years ago, there was no concern about safety at all. We used to operate ammonia from open drums—you open the drum and run. Then when all the pressure came off the drum, you could go back to it. We used to change phosgene in small cylinders. We'd put 200 some odd pounds in a charge so we were changing three cylinders in a shift. We also worked with cyanide, hydrogen chloride, methanol and a chemical named dioxane which I think is similar to dioxin.[1] They used to warn us that it was bad, yet we never took any precautions. It didn't have a bad smell, and you tend to think that if doesn't smell bad it doesn't hurt you. Now I'm at a point in my life where I worry about things. Sixteen years ago, at 30, I was indestructible. Now, I don't even have a smoke alarm in the house that gives off radiation. I looked for the one that has that electric eye in it. I don't smoke anymore. I could feel that I was hurting myself. When I gave up smoking I was working with chemicals, and I heard that smokers develop more troubles with chemicals than those who don't smoke. At the beginning, nothing could bother me. I could handle all these chemicals with no ill effects. Now I wouldn't go into a section that was dirty. If something did break down, it would really have to be cleaned up before I'd go in. But some poor guy has to go in and clean it up. At 30 I didn't worry. At 46 I worry a lot.

### Stuart, mold maker, glass factory

When you're grinding molds, you hold the grinder in your right hand and the wheel spins upward right into your face. It's not too bad on the bench because you can stand back. But when you're grinding neck rings, your face is about five inches away from the piece you're working on and you have to be that close to see. I've done this for three years. Some guys do it intermittently, but I've spent considerable time on the dusty stuff. The nickel scares me because of the powder. You're grinding it off and it's coming right in your face. It's being ground off by aluminum oxide wheels.

The insurance company recommended we have a chest X-ray and a pulmonary function test. So the company arranged for physicals for people over 35. I'm 33 and have worked there 15 years but I can't get the physical. I'm going to push for it because it will help my peace of mind. I think a lot about my health. Call it maturity, as you grow older your values change. Ten years ago I didn't really care. I liked the work. Had a good time. Got more money than I knew what to do with. Spent it like a fool.

### Bill, fire fighter

They're going to do a cursory physical follow-up on the firemen who went to the PCB fire. If they can identify the physical effects, that's going to satisfy them. But there's never been any psychological follow-up. There are people who have gone through this whole thing suffering inside, but the state doesn't want to deal with that. They don't like to deal with "weak" people. But mental health is just as important as phys-

---

1. Dioxane is, in fact, not related to dioxin.

ical health. We know what Agent Orange did. That's where the problem is. It's that time bomb that might be ticking inside you. I spent six months in therapy after Vietnam. You have people that truly feel they have physiological and psychological problems that aren't being dealt with. There's a feeling of betrayal and bullshit.

I'll tell you something, it's a feeling that'll never go away. You know, you can wake up and not have any fears, but when somebody says there's a chance, and nobody will tell you there isn't a chance, it's something that you have to deal with on a daily basis. But they're never going to do anything for those people who can't deal with that. I talked with this doctor who's doing the follow-up. I told him, "It's a joke. You haven't done a damn thing for these people psychologically. They have been exposed to all of this media hype and everything else, and all you're worried about is getting the blood work done. You're asking them if they have these physical problems that nobody would admit they had, or that are so general everybody suffers from them."

I listened to one guy who had spent a few minutes in the building, not even at the fire point: headaches, diarrhea, dark stools, can't sleep, the whole shooting match. He's got a problem. I've heard people worry because their wives have since become pregnant. People complain of headaches, trouble sleeping, trouble not being able to stay awake. Every little patch of dry skin ends up, "Geez, I might be getting something." They have the idea that this thing could happen again at any time so they spend too much time looking for it. With the news about the building in the paper all the time, it's hard for a day to go by without talking about it. But the guys take their pain and everything inward. That's just the way we were brought up. I mean, you're dealing with a person's mind.

---

### "I Worry about the Unknown"

Anxieties in part reflect the uncertainties about the unknown health effects associated with working with chemicals. These uncertainties are compounded by inadequate information. People are often confused about the substances they work with; they are bewildered by conflicting information and unsure of whom to trust. Uncertainties are especially discomforting when workers realize that smells are not an adequate guide to identifying potential dangers. Not knowing what is dangerous and what is not is a source of profound distress. The labels they read, often as the only source of information, hardly help. Reading warnings in encapsulated form contributes to anxiety among those with no recourse, no means of control.

*Bob, fire fighter*

As long as we keep our self-contained breathing equipment on, we're pretty safe as far as breathing in toxic chemicals, unless we get into something that can be absorbed through skin. Normally, I feel pretty secure with the breathing equipment on. I don't worry about that part of the job too much. It's just the unknown; I don't know if you've ever been in a fire, but you can't see anything. You're alone, crawling around, and you just don't know what you'll crawl into.

*Eric, sculptor, self-employed*

When I paint a sculpture with acrylic, oil paints, or shellac, my face is very close to it and I'm inhaling a lot of stuff. I wonder what it is. I'd like to know exactly how dangerous it is. I don't like the label to just say it's dangerous because you don't get any real sense about it. I'm breathing in a lot of turpentine and I don't know how dangerous turpentine is. I know some painters who can't use turpentine anymore because it's really bad. They might have particular sensitivity to it, or it might be dangerous to everyone. I wash the brushes in lacquer thinner. And sometimes, if I strip the patina off, I use acetone. I've heard mixed things about acetone. Some think it's not harmful and some think it is. I don't know. I'd like to see it explained. I'd like to hear the story of how molecules travel through the skin, into the blood, into the kidney, and what they do in the kidney. I read something in *American Artist* magazine about a group of art material manufacturers that had a voluntary program to label things. I would sooner buy a product that has good labeling than something that doesn't. I'm pretty careful with paints. I wash my hands but I don't know how well I have to wash them. Can you just wash with soap and water or do you have to scrub for 15 minutes using solvent? Is it better to use solvents on your hands and deal with that risk to get a little paint off, or is the solvent doing more damage than the paint?

*Elise, laboratory technician, research institute*

I used to be more nervous when I didn't understand much about carcinogens. I think if I'd known then what I know now, I would have been less careful. I get real curious when something says it's a carcinogen. I go and look up the evidence and consider the potency. I understand the risks now and probably would still be careful, just because I'm a carefullish person. But I was much more afraid of it then than I would be now because I know what it means. I understand the nature of risks and know that every cigarette I have totally outweighs whatever I could get making up a few micrograms of this thing that says, "Shown to cause in cancer in animals." Isn't that awful? Knowledge breeds contempt.

*Rose, pill coater, pharmaceutical plant*

One thing that we work with is a soapy kind of a material—a soapy powder. When you dump it all the stuff comes up in your face and you could really choke to death. It's awful. We're supposed to wear respirators, masks, gloves and everything, but I really don't know what it does, that's the whole thing. That's why it's so bad. I don't really know enough.

---

**The Long Term**

The possible long-term effects of chemical exposure are a source of bewilderment and vague anxiety. It preoccupies some workers, but others find it hard to dwell on a problem "that doesn't happen then and there" but might turn up 20 years in the future. Many feel that their primary worries are acute: collecting a paycheck, feeding their families, avoiding disabling injuries. It is difficult to be concerned about unknown and uncertain future effects when such immediate concerns are pressing.

---

*Earl, landscape supervisor, botanical garden*

We're all educated about the symptoms of the different kinds of pesticides. We spray in the buddy system so that we can keep an eye on each other if anyone begins to show symptoms. But, of course, in the back of the mind of everybody who uses these chemicals, including me, are the long-range effects of our frequent mild contacts. The fact that you don't fall down and foam at the mouth and die right there may not be all there is to it, and we're aware of that. We wonder what it does to us. We get lots of literature on, say, Benomyl, and it shows all the things it does to your genes. I don't mean your blue jeans. So, we wonder if, in the long run, the harm may be more than we're aware of.

The man who trained me didn't use much protection. He used just about everything around, going back even to the banned ones, like DDT. He has problems, but there's no way of knowing whether or not they're from work. When you get older, things happen, but you don't know what causes them. That's part of the fear. It's impossible to know. I'm frankly afraid of what might happen. Maybe I'll get cancer and die 20 years sooner than I might otherwise live to be.

*Ken, electrician, chemical plant*

It's hard to think about long-term risks, about something that's going to affect you in 20 years, some cancer-causing thing. You can have a sweet-smelling chemical like toluol and you don't want to know that it's hurting you. That's not something that's normally discussed. A health study brought out the cancer rate but then it was quickly forgotten.

### Rich, orchard worker

Almost every job has risks, but there are some things at the orchard that I refuse to do, like going up in a tree with a chain saw. It's too easy to take off a leg or an arm. Working with pesticides is different. It's a danger you don't really see, a danger that doesn't happen right there and then. It's much easier to ignore, less threatening, less scary. As much as I try to be aware of the long-term dangers of using chemicals, they somehow don't seem as imminent as cutting off a leg.

### Bob, fire fighter

I'm 50 years old and I work with fellows who are between 20 and 30. Twenty years down the road is certainly not the same to me as it is to a 25-year-old kid. Twenty years down the road I'll be 70; 20 years down the road he'll be in his prime. So these fellows are worried. What if 25 years down the road they come down with some kind of exotic cancer? The government never learned to deal with the problems of Agent Orange veterans, and they're not going to deal with the problems of the firemen.

---

### "Will I Have Normal Kids?"

During the past five years researchers have begun to examine reproductive problems resulting from workplace hazards. They have found that sterility and genetic disease are associated with a number of substances, including heavy metals such as lead and mercury, various pesticides, and the alkylating chemicals used in drugs. These reproductive problems are a source of growing anxiety among workers.

Those who had direct experiences with reproductive or genetic problems, or who had thought about the issue (mainly men and women of childbearing age), worry that their health problems may extend to their families and their ability to bear normal children. They are also concerned about management practices that are intended to minimize reproductive hazards. This usually involves banning women of reproductive age from working in jobs that might involve exposure to substances associated with reproductive problems.[2] Some of the workers we interviewed insist that reproductive hazards are not restricted to women, and

2. In some widely publicized cases, women have had themselves surgically sterilized in order to keep their jobs. Several companies are in court over their policies that effectively result in such practices.

that, if women are to be excluded from certain jobs, men should be excluded as well.

Concerns about reproductive problems are intense; the idea that work may put food on the table but bring harm to one's children may be the deepest of all fears. Workers' statements suggest the potential volatility of this issue as awareness of reproductive problems increases.

---

### Laura, filter cleaner, pharmaceutical plant

When they originally found out about methotrexate, nine women were removed from two or three different areas involved in the heavy weighing, coating, and compression of the powders. These are predominantly higher-paying male jobs. There's also a light-weighing job with all women where they use methotrexate. Here they got Engineering to make a glove box and didn't ban women from the job.

I don't think it's right to ban women. I don't think it's right because most of the jobs are male-oriented, and they are higher-paid jobs. There are women who are trying to make more money because they're supporting themselves or their families. I think that banning women is a cop-out. Maybe the probability of damage to the cells is greater when I'm pregnant, but that's only nine months. I could have been working there for 10 years, and if you're going to tell me that hasn't affected me in some way, I think that's foolish. When they take women of childbearing potential out of the workplace, I believe there's a hazard there for adult males as well as females. But they don't take a sperm count to see if you have dropped below fertility level or if you have abnormal sperm. They don't try to give you tests for cancer. One of the biggest problems is with the men. They aren't going to say that they're impotent, or that they're sterile, or that they can't have another baby, or even that they're trying. The girls are more open with each other. They'll say, "Gee, I got my period twice this month," or "I didn't get it," or "I'm having a problem," or "I'm going to the doctor," and this and that. They'll tell you if they had a miscarriage or something, while guys are much more protective of themselves. They think it takes away from them as men.

A woman at another plant who had a miscarriage told me that five people in her husband's department had lost their kids. These men handle methotrexate and a lot of other material in the raw form. Their wives got pregnant over about a five-month period, and they all had miscarriages. But none of these men would come out and talk about it. She had warned her husband, but he was afraid he would lose his job. So he lost his kid instead. It doesn't make sense to me, but somehow it makes sense to him. The biggest problem is you can't get people to get out there to protest. They have to have the baby that's deformed, sick as that sounds. A miscarriage is something that can be forgotten. A deformed child that gets hurt, that needs medical care, that could die, that's very real. I don't really know what will happen, but I think that something is going to happen that'll be pretty nasty once people realize this thing.

### Sid, chemical operator, chemical plant

We had a meeting with some researchers who spoke a lot about reproductive problems. A lot of wives came. I brought my fiancée. She's extremely worried about it. It's something to think about, but you just can't quit because of that reason. Anyhow, it's not in the nature of men to sit around a table and talk about this kind of problem. But whenever people bring their wives or girl friends, everybody starts to open up a little bit. You find a lot of common problems.

### Joe, laboratory assistant, chemical plant

I worry about what I'm working with especially when all the stuff comes down about fetus damage and women who are having miscarriages. That shakes you up. They now won't let women work in certain areas and that's good, but then everyone should be excluded from something like that. They should find something else to use, an alternative or a safer way to use it. I mean, if they know it affects women and they won't let women work with it, how do they know it doesn't affect me?

Down in the lab we use one chemical that women aren't allowed to handle. So if one of the girls has to use it, she gets one of the guys to pour it and mix it for her. I've refused. If it's going to bother her, it's going to bother me. They can just get someone else to do it.

### Les, furniture restorer, self-employed

I've always worried about the possibility of a malformed child. I don't think I could handle a malformed child. I'd have a hard time with physical deformities, but it would be worse with a child that was mentally no go. Today the screening procedures are so advanced that it is possible to have a mistake aborted, as long as these jokers, these right-to-lifers, don't fuck it up. I'd feel especially concerned if a child was malformed because of my work. But, at this stage, there's nothing I can do about it, apart from stopping doing my job.

### Mike, photo lab processor, blueprint company

There was a girl who worked here a short time over in the other side of the building in offset, making all these cards by the thousands. They have some screwy little chemicals over there. She just had a baby, and of course nobody mentions it, but that baby has a slightly deformed head. Now, how or why, this is anybody's guess. We're obviously not going to pin it on the company, but it happened, it's there. We don't know anything about genetics or what it takes to screw up a baby, but all I know is that she's got a reality. She's got a baby that's bad news.

### Penny, laboratory technician, health department

Some high official in the government said that women of child-bearing age were not allowed to work on the dioxin samples because of the spontaneous abortions and childbirth problems they had at Love Canal. My supervisor pulled me off and said, "Don't do anything with the samples." Then no one wanted to touch the samples. The guys started complaining, "What about us men of child-bearing age?" "This isn't right." If you take away all the people of child-bearing age, you have to ask for volunteers. Who wants to do it? It really got to be a hassle, and the higher-ups in charge of everything came up with an idea. The procedures have to be so safe that they would threaten no one. Then my supervisor, who was, like, really against me going near the samples, said, "Okay, everything's safe now." Actually, I think everything was the same. I didn't like the fact that one day they pulled me off and said, "It's not safe for women of child-bearing age," and then all of a sudden, "It's safe for everyone."

### Dick, granulator, pharmaceutical plant

Methotrexate is known to cause fetal damage and they didn't want to take chances with the women, so they banned them from working with methotrexate and acetazolamide. But it also has adverse reactions on men. It supposedly lowers the sperm count. But they didn't want to talk about it. They didn't want to give us the actual information that they had on human testing. We're still required to work with it. A lot of people don't think about it. But if it affects the women, then why not the men? My fellow workers are saying, "Hey, why should only women be banned? We're being exposed to it too, you know? There's problems with us, too!"

### Rose, pill coater, pharmaceutical plant

When I was down on the packaging floor, we worked with tablets already coated. The only powdery thing we ever worked with was acetazolamide. There's a trexate (methotrexate) which I'm not allowed to work on because I'm of childbearing age. I don't even go near it. These kinds of chemicals are what really scare me. They once tried to get me to work on methotrexate in the filling room. It was like, no way. I'm not going in there. You can't make me go in there. People said you can't refuse a job, but to me, if it's going to hurt me I'm not going to do it. They wouldn't accept the answer NO, so I just got somebody to replace me. I talked to a friend who was older, and she went instead. These reproductive risks are the only thing that really bother me. I've hassled the company about the younger guys going in. To me if it can hurt a girl, it can hurt a guy. What's the difference? You never know what it can do to you. I feel that young women definitely should be excluded from working with methotrexate because there are enough older women who can work with things like that. It's the same thing with the young guys, but they have no choice. They can't say no. But if a guy has been working on it for four or five years straight and his wife has a baby that's deformed, it's got to be him. There's nothing in writing that says it's going to make you sterile or affect your child, so you can't refuse to do your job. Some of the guys worry about it. Some of the guys don't care. It's like, "Well, what can I do?" They talk to me, but they can't do anything. They can wear bunny suits and take all the precautions, but it's still scary.

### Earl, landscape supervisor, botanical garden

I want to have children. My wife and I are getting to the age where we can't put it off much longer. Suppose because of the contact I've had for 10 years with chemicals, I have a deformed child. I don't know whether there will be any such effect. There's no proof for it, but there are certainly lots of things that make me wary and I'm, frankly, afraid.

### Don, railroad conductor

When the train goes into a terminal, the brakeman has to go out and throw the switch. The terminals are sprayed with dioxin. It's all over the rails, the roadbeds, and the switch handles. So when you pick up the bar to throw the switch, or get out to release the brakes, you're exposed. It's thick, oily, like a white cake all over the ground. I knew that it was a weed killer, but not what chemical it was. I found out it was dioxin and that there was something wrong with it from some articles in the newspaper. I found out that dioxin could be linked to certain types of cancer. Being my daughter had cancer, I got interested.

The railroad insists that they haven't used that type of chemical for four years. But for the 11 years I've worked on the railroads, they've been using some type of weed sprayer. They have to, to keep the weeds off the roadbeds and the yards.

It's scary to think that I could have brought the stuff home on my pants or shoes or clothes, and maybe I threw them in a pile in the corner, and my daughter crawled over them. Or maybe I picked her up and it was on my coat. Something like that could have possibly caused her cancer. I don't know if it's a guilt feeling on my part, it's just a mixed emotion type feeling. It's hard to explain. I hope that wasn't the reason. You read so much about it. It's a very hard feeling to describe.

# 4    What's to Blame?

Industry and labor are inclined to have very different views about responsibility for exposure to toxic substances in the workplace. Placing the blame for health and safety problems is important because culpability serves to guide industrial practices and public policy, and to define responsibility for compensatory and remedial measures.

Before 1970 the organizations which dominated safety concerns in this country, including the National Safety Council, the American National Standards Institute, and the American Occupational Medicine Association, tended to emphasize the responsibility of the worker for most health and safety problems. Many people still assume that workers rather than equipment or management practices are at fault in most accidents, that health hazards have been substantially controlled, and that management adequately tests substances for potential risk before exposing workers. If workers are exposed to toxic substances, it is often blamed on carelessness, failure to follow correct procedures, or unwillingness to use precautions. If workers do get sick, it is often attributed to personal habits such as smoking or, more recently, to genetic predisposition.

Such views have long dominated the literature on occupational health and safety and are often accepted as facts rather than representations of an industrial perspective. An article on occupational psychiatry in the 1966 *American Handbook of Psychiatry* illustrates the translation of industry's view into general belief. It attributes the pulmonary insufficiency ("pneumoconiosis," "emphysema," "chronic bronchitis") of workers in dusty conditions to depressive reaction, anxiety reaction, and psychophysiological reaction.

Another view began to emerge into public discourse in the late 1960s as worker health advocates began to attribute occupational illness to problems within the control of management. Assuming a basic right to health, they argued that work should be made safe for all employees. While the reality often lies somewhere between, views toward responsibility often remain polarized.

In this chapter, workers describe problems—of workplace design, of supervision, of knowledge, and of corporate goals—that they believe contribute to the risks they encounter on the job.

### Workplace Design

In seeking the sources of chemical hazards, many workers point to inadequate ventilation, poor design of equipment, and careless maintenance. They perceive an overreliance on makeshift measures to keep equipment in a safe condition, and they judge these measures inadequate to protect their health. Some suggest that the source of the problem lies in the failure to either maintain or replace obsolete equipment: "Things don't get done until somebody gets hurt." They want proper ventilation and maintenance, and a production schedule which would allow them to follow safe practices.

---

### Joe, laboratory assistant, chemical plant

Just about everything's wrong. They have exhaust fans on the ceilings in every section and I don't think a third of them work. They have elephant trunk hoses, and most of those don't work. Everything is enclosed. The only place that isn't enclosed is where they work with cyanide. There they have open sides so the wind can blow everything out. The hoods in the lab are pathetic.[1] It's such a small lab that they have equipment stuck in the hoods so you just don't have enough room. The sink drains are in back, so if you have to pour something down the drain you have to go inside the hood, and once you're inside there's no sense in having the hood. When we complain the reply is, "Engineering is looking at it." After a few years, Engineering does look at it, but then it requires parts and these can be on order for a year and a half. When they get the parts in, it takes them six months to get it together, and when they do get it together they have to take it apart and do it again because they don't do it right. The company can see no reason why anything should be shut down if we need that product: "A piece of tape will fix

---

1. Laboratory hoods are ventilation units designed to enclose toxic materials during experiments.

the leak until we can get it done. Just be careful. Don't walk under it, walk around it." That's pretty much the way they run the plant.

### Mike, photo lab processor, blueprint company

The room we work in has to be dark, and usually it's enclosed. Ventilation is very poor because the building is old and vents were never designed in. We complain about it, but they don't care. These are closed rooms because they're darkrooms; there's no windows, no escape hatches, no nothing, and a lot of these are one-way rooms with one entrance. Fire extinguishers? There aren't any, except in the main hallway. So if you're in one of these big rooms and you have a fire, you're up the creek. No sprinkler system, no smoke alarms, no emergency lighting. These are darkrooms. I mean, you can't get out because you don't know where you are. One day, I had nothing to do, so I opened up the ceiling panel to see if I could crawl through someplace. There's only a two-foot space and you really couldn't go anywhere, so you're pretty well trapped. If it was a slow fire—we got chemicals back there, I have no idea what these would do if they burned.

### Sheila, laboratory technician, research institute

We were working in the fume hoods. Well, not always . . . a lot of the time, the fume hoods didn't work so well. The problem was that the stuff that went into the fume hoods blew out of the building on the south side. When the wind came from the south it would throw the fumes back into the clean air intakes through the building. People would say to me, "Come down on the second floor and smell this stuff." I would have to identify what it was and then go through the building trying to locate who had been using what. Once, a lot of management people came in and tested the system. They found that the duct work was defective but within legal limits. Just. The size of the ducts themselves were bigger than they were supposed to be but within specifications, and the dampers were smaller than they were supposed to be but at the extreme of the acceptable size. There were clearly gaps, but because they were within specifications nobody took responsibility and they decided to let it go.

They also didn't design the building with a place to store pesticides, so they stored all the chemicals together in the same place. I went to the Safety Committee and told them we had a problem. "What happens if you've got a fire and the fire fighters come in and they're not notified that there are pesticides in that room? They should know that for their own personal protection." I wanted to put a sign on that door, but the chairman of the committee told me, "You can't put a sign on the door. It's illegal to store these things in the same place. We don't want to be blatant about it, we just want to do it."

*Daniel, chemical operator, chemical plant*

The equipment here varies. In certain sections they got up-to-date equipment; other stuff looks like it goes back to 1909. It's old. That's why they have a lot of problems. They push the men, "Come on, we've got problems, let's go, let's go." But you can't do that if your equipment ain't running right. What are you gonna do? Right away they blame it on the operator. I don't want to hear that. When I run across a little problem I can usually take care of it. But if something's really wrong it goes through a lot of places before it gets fixed. Things don't get done until somebody gets hurt or killed. Then they say, "We've got to fix this; somebody just got hurt over there and it don't look right. If OSHA comes in, they're gonna close our doors."

*Sue, laboratory technician, university*

The main thing that bothers me is that people eat lunch right at their lab tables because there's no place else to eat. Since we only have 30-minute lunch breaks, we don't go out because we would spend half the time waiting in line. If you know what a lab table looks like . . . I work with all these different chemicals and usually wipe the table up with paper towels before eating. But frequently someone eats lunch on my lab desk while I'm still working. One day, the supervisor of the lab told me that he had gotten sick. He asked me what chemicals I had been working with, because he happened to be eating lunch sitting next to me the day before.

*Jack, nurse, hospital*

The hospital where I work is a very new facility and well constructed. In the operating room the machines that they use for anesthetic gases have rebreathing circuits. In the old hospital the gases were vented directly into the operating room. When the patient exhaled, all those gases just went out into the air. Now they have filtering systems on the machines that are all rebreathed. That is, the gases that are still potent are recirculated right back into the patient, so you don't often smell them.

*Walter, pipe fitter, glass factory*

It's not a bad company. I'm not going to say they're terrible. They try hard. Basically it's one of the better companies I've worked for. But I think they miss a lot of things that they could do. Like the stannic chloride system. They tried to improve it, but it still plugs up and puts the stannic chloride on the floor. I think they could get a better system. Sometimes it gets pretty smoky in there, but they tell me it's 100 percent better than it was 20 years ago. They say when you used to walk in you couldn't see two machines away. Now you can pretty well see down through there even though there's heavy smoke. Things are awfully greasy. It's nothing when we go to fix a machine to find black grease an inch thick on the pipe. They finally put in dust collectors. We argued for six or seven years to get them.

---

### Our Bosses

Many people blame their problems on the system of supervision and its preoccupation with production. Just as factory workers suggest that production takes priority over health, so some laboratory technicians feel that research takes priority over people. Some workers see the problem of supervision in terms of style: the absence of role models and appropriate guidelines, the distance between management and actual operations on the shop floor, or the arrogance and carelessness of their supervisors. Others question the legitimacy of supervisory authority because of incompetence, or they resent the hierarchical arrangement itself. Some see their immediate supervisors as simply reflecting the demands of their own bosses; caught in the middle, they neglect health and safety to protect their own status. Others believe that higher management is ignorant of conditions on the shop floor and that if they really knew what was going on, they would insist upon change.

---

### Nick, chemical operator, chemical plant

The biggest problem this plant has is that anybody with a degree thinks they're above the men on the bottom rung. There is no communication whatsoever. They have no respect for their chemical operators. They think we're a bunch of dummies because we don't have a degree. But we're as smart as they are. Ain't anybody knows that equipment any better than the man who runs it. They just aren't willing to listen. For some reason, when they get that degree they look down on the people who are underneath them. You can be a brilliant person, but if you have no common sense, you're the dummy as far as I'm concerned. That's the way a lot of them are, especially the chemists and engineers. One man down there, I guess he was third or fourth in his class at M.I.T. I was working on a project and he told me we were going to do such and such. I told him, that equipment just ain't gonna work, but he went by the book. Sure enough, it didn't work. I called him up on the phone and said, "You know what you can do with that book." The books give you a concept, but the concepts are not always true.

### Paula, electron microscopist, pharmaceutical plant

My supervisor is really nice, very considerate, and aware of the dangers of the materials we work with. He insisted on having the hood put in; he is always real careful about making sure I wear gloves. The only thing is he doesn't want to waste anything. He's Indian, not American. Americans throw everything out and don't think much about disposable things, but he wants to wash and reuse everything. That's bad because you'd get this plastic stuff and have to soak it in Epon to reuse it. When you're working with Epon, and acetone and all this junk, and changing the solutions to soak it, you inevitably get it on you.

### Bess, diffusion analyst, manufacturing plant

Whenever safety and health are talked about it's obvious that the front office is ignorant about the things that go on. I think they sincerely believe that things are correct. The negligence is at a lower level. They pay Facilities to do the job correctly and take it for granted that it's being done. But then they start too many jobs at once, and move the men from one job to another before the first one is finished. One of the stewards went to her foreman to complain about an odor in her area, and he said, "Well, if you want a job, you have to take the odors." That was a very negligent answer because the odors are not perfume. They're either solvents or acids and you don't fool around. When you go to a foreman, you shouldn't get an answer like that. So I think the problem is at a lower level than the front office.

### Tom, print machine operator, university

There was a problem with the disposal and storage of chemicals. It was especially bad for people who smoked. So instead of trying to solve the problem, they eliminated the smoking, which was more or less a form of punishment. That's a blame-the-worker approach; it's a way of saying, "O.K., you bad people, this is what you get for speaking up." To this day, there is still no smoking allowed in the press room.

### Greg, air-conditioning repairman, university

Each lab is run separately. The supervisors are researchers who have complete autonomy, and some of them feel that regulations are laughable. You know, some of these professors, I honestly believe, are willing to die for their experiments. They are more interested in not disrupting the time schedule of their experiments than in the safety of the workers who fix the stuff.

*Sheila, laboratory technician, research institute*

The attitude of scientists in research labs is that they're really above all of the problems and that anybody who works with them should be grateful. I don't think there's going to be any change in their attitude. They view OSHA, or any kind of regulatory agency or any force that imposes safety regulations, as a real imposition on their lives. They believe that science must go on at all costs. The safety committee feels that these are scientists of renown and that it can't go around and tell these people what to do. The chairman of the Safety Committee was a guy who was chosen for that position on the basis of his complacency. He was a chemist, but he really didn't understand how all of this chemical information related to human beings. And he didn't choose to find out. So, there was no control in the lab. With scientists as supervisors, nobody is really responsible.

*Rich, orchard worker*

The head orchardman had a very cavalier attitude toward pesticides. One of his famous lines was: "I've been using this stuff for 15 years and it hasn't hurt me." The attitude was, "Better get the job done. Don't worry about this stuff." Once I was handed a pesticide in a little bag. It was an experimental chemical, with just a number and no label. I was told to put it in a tank and stand underneath the trees and spray it up. I put a rainsuit on, but I was told I didn't need a respirator; it was harmless stuff that wouldn't hurt me. So I started spraying. I was really getting dripped on quite a bit and decided I'd better check out the stuff. I found the bag which just said to protect your eyes because it could cause an eye injury. Yet my boss just dismissed it with, "Oh, this stuff's not very toxic, it won't hurt you." So there's this casual attitude that comes all the way down from the top. There's no safety consciousness. There's nothing coming down saying that "if in doubt, you should look into it further." The idea is, "If in doubt, ignore it and do the job; that's what you're being paid for."

Things changed after we went through the Certification Program because people found out just how toxic some of these substances are. It scared my supervisor because he learned some of his cavalier attitudes could easily have led to injury. He had never used a respirator because he could never smell the stuff, and he figured if he couldn't smell it it was diluted enough and couldn't bother him. Well, the safety people told him that there'd been a number of deaths attributed to paraquat, and the reason he couldn't smell it was that it was odorless. That shook him up!

### Mike, photo lab processor, blueprint company

Just to tell you what our manager is like, one day an ammonia leak developed. It don't take much ammonia to get your attention! Whoa, get me out of here! I mean, it was dangerous. Ammonia burns your lungs and you can't breathe. Nobody else was concerned, but eventually they started to get out because it was getting worse and worse. We called the vice-president to find out what was happening. The boss of that section insisted, "There's nothing wrong, nothing wrong. I don't smell a thing." Right? We were starting to literally choke. Our bosses were standing six feet away from those tanks trying to look like nothing's wrong, but their eyes were watering. Here they were gasping for breath, and they're trying to say nothing is wrong. They're trying to keep working, hanging around, smoking their cigarettes. It didn't work. They eventually shut the machines down and found the leak.

But this is the way they are. Anything that goes wrong—machine breaks down, box of film gets fogged—in their view it's never the machine's fault, it's always human error. They couldn't care less if you died, except that if you die you should die outside so they don't have to have an ambulance come with a stretcher. This is the way their thinking is. Many people are stupid in assuming that they're going to be protected by their employer, but it seems like it's the other way around—they're gonna be used left and right. You know—who are the bosses anyhow? They're people. They seem to be up on some level. "You're the servants, you just do what we say or we'll zap you." It isn't so much that we're in any danger; it's just we are in a lot more danger because they're not taking safeguards. In cut and dry language, they just don't give a damn.

### James, computer assembler, manufacturing plant

They try to make you feel like the problem is yours alone, even the chemical problems. If you come down with a rash, even if 10 people come down with a rash, it's all your individual problem. It's not the company's fault; it's not the chemicals; you have an allergy. One woman had the whole side of her face swell up after working with a degreaser, and they said it was her own personal allergy. Soon as she left the job it disappeared. One woman lost her larynx. She thought she was coming down with a cold, and the doctor said her larynx looked like a 100-year-old piece of paper. It just kind of dried out from the degreaser she was working with. The manager transferred her to another work area very quickly. When anything happens they tell us, "Don't tell anybody about this problem, we'll take care of you." When they tell us this it's an implied threat. While they pat you, they're sticking you up the ass. They tell you to go through the chain of command at the company if you have a problem. But the manager is expected to stop it right there. Upper management doesn't want the problems to get to them. They don't even want to hear about it.

### Steve, railroad trackman

One engineer who is in charge of a whole department has a parking space right outside his office, with a sign on it that says "Colonel Klinker," referring to the Nazi commandant of the POW camp in that television program, "Hogan's Heroes." I don't find it funny. The space right next to that says "Sergeant Schultz," who I guess is one of the other Nazis. Of course, it's just a joke, but I'm among those who don't think there is such a thing as a joke about Nazis. Anyway, that's the type of mentality that's typical of our supervisors. My own supervisor is different from them. He's a lot less predatory toward his subordinates, and he's much more easygoing. But I still have problems working for him because of the structure of the railroad and the conditions of work, which, in many cases, are beyond his control.

### Rose, pill coater, pharmaceutical plant

One supervisor is excellent when it comes to safety. He is so protective that sometimes he's annoying. He could drive you nuts. But it's good. I'm glad he is that way because maybe if he keeps getting on people's cases they'll start taking care. The other supervisor, he just wants production and doesn't care how he's going to get it. You don't have to wear your mask. If it's going to take you too long to put it on, don't wear it. The supervisor that does care is not the norm. Most of them don't care. As far as they're concerned, it's your problem if you get hurt. It's your responsibility to put the stuff on.

### Jenny, laboratory technician, pharmaceutical plant

The $H_2S$ down there is very bad. The ventilation is the problem. In where the filters are, there's supposed to be ventilation all the time. They're supposed to have it checked regularly, but it hasn't been checked for over a year. You can't stand it. You go home, you've got to wash your hair every night. I walk in my house at night and they tell me I stink. We put in grievances. I've gone to my boss. I've gone to my department head. I've gone to my section head. They all say they're doing something, but they move so slow at doing it. I've never seen a big company move so slow.

*Sally, services technician, hospital*

The people I work with make the job very depressing. And right now we're going through a terrible situation in our department because our department heads are new and they're so insensitive to black people. I've spoken with the directors of Central Services about the autoclave. These people don't know the risk of the things that they're bringing in. The machines are always malfunctioning, and the aerator is not venting like it's supposed to. All these folks are exposed to these faulty gas things day after day.

One morning about two weeks ago, I happened to open the door to the gas sterilizer where we use ethylene oxide. This big gust blew out, you could feel it and smell it. I went to the grievance proceedings and said, "I think those folks are exposed to a lot of ethylene oxide that they have no idea about." I tried to explain to management that I was reading a lot of material on ethylene oxide and that, with my background in chemistry, I was concerned. But management's thing is like, "Look, you stay in your place, this is none of your concern." It's almost as though they're running back to the days of the cotton fields.

---

### Inadequate Knowledge

A substantial portion of blame for unnecessary exposure to chemical hazards is placed on the failure of management to inform workers about existing hazards and to train them in proper handling procedures. But some people also complain about co-workers who are careless and fail to heed warnings and instructions. They call for more enforcement of rules, and suggest that workers who are intimately familiar with equipment and processes should themselves have greater responsibility for implementing appropriate health and safety measures on the shop floor.

---

*James, computer assembler, manufacturing plant*

They gave us a chemical precaution sheet with the hazard level of each chemical. Some say it will affect your liver or kidneys, but it's never explained. It's tossed to people who have maybe just a high school education, and there's a word on there like trichloroethylene. If you've never heard of that word before, you wouldn't even know how to pronounce it, let alone understand it. And the threshold limits are all Greek to people. It's never really taught that this is how to decide when you are being overexposed. They're fulfilling their obligation by giving you the information, but they're not telling you how to understand it.

### Rich, orchard worker

When new people come in, they don't necessarily know what to do. A high school kid was working for us, and he was sent down to the tractor shed and told to measure out the Guthione, five pounds in each paper bag. He was told to wear a respirator. So the kid went down there and was working merrily away. I came walking down to get something and I happened to glance at him. Yes he was wearing a respirator but he was also wearing a tee-shirt. He had rubber gloves that came up to his wrists, and that was it. His arms were coated with the Guthione, which can be absorbed through your skin, and here's this kid saturated with it, this powder all over him. I freaked, sent him back out to get a shower. That's the kind of thing that happens. He wasn't trained, he wasn't even shown properly how to do it. They said "Wear a respirator" and assumed that he would then know that he should wear these other things too and that he shouldn't get it on his skin. Well, he didn't know that. He assumed that you just weren't supposed to breathe it.

### Lisa, laboratory technician, university

No one ever told me what chemicals were bad or good or dangerous. I sort of picked it up from comments that people made. One chemical—it was toluene—that I routinely worked with was carcinogenic[2] or was very bad to inhale. It should be used only under a hood. I found this out months after I had started using it, and then I did use it under a hood.

### Jocelyn, secretary, museum

I have to remind myself that it's not just in my head, that there's a physical thing going on. It's a constant battle; we have to talk about it all the time because the minute we stop talking about it everyone just seems to slide back into thinking that "oh, we must just be tired," or "oh, we must be depressed." I don't know if it's just the way we're socialized or if it's coming from certain people at work, but I think there's some kind of insidious pressure to feel like you're exaggerating. I feel these things, but then I think that I don't really have any hard evidence to support that the chemical could be causing really dangerous changes in my body, so why should I be talking about it? Maybe I *am* just getting hysterical. The first thing that occurs to you is that your fear is making you make up things. I have to say to myself over and over, "Seventeen

2. Toluene is not acknowledged as a carcinogen.

people out of forty have rashes''; I have to repeat these numbers to legitimize my complaints. If I didn't know other people with those complaints I'm sure that I would have convinced myself that I was causing it by being just nervous or exhausted. People tend to blame themselves.

### Bob, fire fighter

It's just the uncertainties of the job, you just don't know what you'll run up against now. When I first came on the job, there were many dangers but they were apparent. You could fall through a hole in the floor, or have the roof cave in on you, or get a back draft when you open the door of a burning room. There have always been a million hazards, but the job today just isn't what it was. What's changed? Look around. Polyesters, vinyls, plastics, nylons—all the things that are in the home, the workplace. You're not just breathing old carbon monoxide anymore, not that that didn't kill you bad enough. When I went on the force, you'd go to a house and find a class A fire with wood and paper and cloth. Today you go in and God only knows what you find. Everyday materials, like foam cushions, release hydrochloric acid. You get everything you can imagine. They don't really know what this stuff does to people. I kind of feel that fire fighters are being used as guinea pigs.

### Vivian, laboratory technician, research institute

The chemicals were generally labeled, but you had to trust that if people were making up their own solutions they were going to mark down on the beaker exactly what was in there so you would know. Sometimes, there were unlabeled solutions just lying around that people might have poured into a small container just to use for one experiment. You wouldn't know what to do with it. Often when they were done, they'd forget to dispose of it. You would just have to use your personal judgment on these things because you really weren't sure whether or not it was dangerous.

### Walter, pipe fitter, glass factory

The biggest problem is they come out with a bunch of numbers in language we don't understand. If they'd break it down in plain everyday langauge, ''Hey, this is what it is, this is what it can do to you,'' they'd get the people to take precautions. Let's face it. If somebody tells you that something is going to hurt you or kill you, you're going to pay attention. But if they give you a bunch of figures and numbers—that looks impressive but it doesn't mean a thing to me. That's what happens. I think sometimes they really feel the people are more educated than they are. They aren't ignorant but they don't have any knowledge in that line. They are smart people but some of these things, like anhydrous stannic chloride, if you ask a person, "What's anhydrous?" they wouldn't know. I know it has something to do with water. But I couldn't tell you exactly. The same way with titanium chloride. They come out with a bunch of stuff that doesn't really tell you what to do if exposure is suspected. They say, "Wait for symptoms to develop." Well, what are the symptoms? How do you know when something's going to develop when you don't know what the something is? It's that simple. There's not enough information.

### *Eve, sorter, manufacturing plant*

They've trained us to work with the acids. They've showed us movies on how to use everything. Yet the people still go ahead and work as dangerous as ever. All that general training doesn't substitute for specific warnings, and they never have warnings on things. When I first worked there they used to have signs that you always put at your sink, like—ACIDS IN THIS SINK, DANGER. Well, they've let that go down the drain. We keep trying to tell the company that a lot of their information gets lost in the shuffle. That's why every so many months they should have those safety courses. Maybe it's going to cost them a few dollars, but they have to give them over and over again because you keep getting new people from the bumping procedure, and the new people just don't understand what to do.

### *Kitty, industrial painter, university*

When I came in as an apprentice, I didn't know what these things were that I was working with. It's funny how few of the journeymen know what the dangers are that you can run up against, what the chemical will do to you or what's really in it. If you refuse to use it, you might get laid off, fired, or suspended, or at the least the boss will remember who you are. So most guys don't want to rock the boat. I guess they figure ignorance is bliss.

My first day on the job, we did a large construction project. I was just so excited to be there. It was very strange for all the guys because they had never seen a little blonde girl walk on and say she was a painter. They had me rolling epoxy on the barn walls. We had two colors—bullshit brown and calfshit yellow—so the paint wouldn't show how much dirt there was on it. I wore a short sleeved tee-shirt, and the first day on the job I was awful sloppy and came out just covered from my fingernails halfway up my shoulder. I didn't think anything of it. Paint, who cares, just paint. Then one of the guys mentioned that we're supposed to get extra pay for epoxy. He didn't tell me that it was because it was hazardous. Since then I've found out some pretty horrendous things about epoxy; it makes you bleed through your pores, it burns your lungs, burns your eyes, gives you ear infections, and, in short, can kill you. Here I was for weeks getting covered with this stuff, and it absorbs through your skin. I guess I didn't get sick because we were working in a large barn with lots of ventilation. If I had known what it was, I might have worried or at least tried to keep it off me a little bit more. I was so concerned with working hard to prove myself to the guys that I didn't think of anything else. Somebody could have told me; they could have said something.

---

### The Profit Motive

Generally, American workers in private industry have shown relatively little class awareness. Except for periods of militancy during organizing drives and major strikes, they have seldom expressed their problems in political terms.[3] In this spirit, few of the people we talked to identify their daily interests as directly opposed to those of management and the owners of capital. As one man told us, "If they don't make a profit, we don't have a job." The issue of health, however, is sufficiently powerful to elicit a political response. A remarkable number of workers explicitly blame the conditions which expose them to risks on the "profit motive," and they suggest that the corporate drive for profits is fundamentally at odds with concern for workers' health.

---

*Nick, chemical operator, chemical plant*

The safety administrator's got some peculiar ideas. His big bugaboo is dirt on the floor of the reception areas. "Stop everything and get the floors washed." But when it comes to replacing a valve or pump or installing better ventilation—anything that costs money—nothing gets done. If it's going to cost them money, they don't care. The problems come when you have to have down time for safety purposes. They claim "safety is their most important product," but when it comes down to actually doing something, it's not true. They wait until something goes drastically wrong before they'll do anything. They have no preventative maintenance program. They put out the product to make money; that's the bottom line.

*Mary, housewife, wife of a railroad conductor*

I guess I'm cynical about things from our past experience. It's going to be a very long time before I would trust my employer or the government. The railroad would never tell the truth. Do you honestly

---

3. In this sense the American labor movement has developed quite differently from its counterparts in European countries.

think you could believe them? They're probably still using the same old stuff on the tracks. If dioxin's going to alleviate the weeds, and it's the cheapest way to do it, then they're going to use dioxin. They're not worried about individual lives, they're worried about money. I think they'd go to any extreme to save money, regardless of whose life it endangered. When you question the people in the management, they ask, "Well, do you really know what happened?" I'd like to take a few of these people and bring them up to the hospital where they give kids chemotherapy, or cut their legs off and do other gross things. Then let them tell me that dioxin or whatever else it is they're using isn't going to hurt. I doubt that these products will ever be taken off the market unless the president of Dow Chemical loses a kid.

### Sandy, rigger, chemical plant

The company cares nothing about its employees, or even its supervisors. It's just a board of directors who only care about profits for their stockholders. They will put up the facade of being safety-conscious, but the reality of working conditions—that is of little concern, because fixing them would reduce profits. I personally believe that the rich, the powerful, the large corporations, take advantage of the workers. We've become dehumanized, subject to machine-speed theories, as if we're mice in a maze. We've been toyed with, played with, and symbolically given compensation in the form of paychecks every Wednesday, with little or no regard for how long we'll enjoy the paycheck or for the economic hardship on our wives and our children if we were to die of an occupational disease. It's of little concern to them, it's of great concern to us.

### Laura, filter cleaner, pharmaceutical plant

The company has a big safety program, but it's very superficial—they'd rather put the burden on the employee. Think safety, wear your hardhat, do this, do that. Everything is the employee's fault. To me it's really a protective device for insurance purposes; the bigger the safety program, the smaller their insurance premiums. Everybody in the safety department wears a hard hat, even when they're on the john. They're never without it. I'm sure that's a condition of employment. They also have floor mats in front of all the elevators saying, "SAFETY," but you always trip over them—kerplunk! It's a real big joke all over the plant, and they've removed some of them already. It's all a farce.

The pollution problem is related. People who live near here are very concerned. They called the union local about the smells and the stuff in the brook where their children play. I don't blame them because the company is not very careful about where they're dumping toxic wastes. They just want their profit and nothing else matters.

*James, computer assembler, manufacturing plant*

I can imagine a factory that's building health and safety equipment but at the same time hurting the workers in the plant. I can imagine a company that's building new air-quality monitors but having their workers dip the stuff into trichlor (trichloroethylene) to degrease it before it's put together. It would not seem absurd to me at all.

*Walter, pipe fitter, glass factory*

With the economic situation getting worse, I don't like to see the company fined. You don't want to close the plant down. I think they are like a lot of other companies. They're trying to make a profit and get along so they just let things go. The bottom line is, if factories don't make a profit, I don't have a job.

*Ted, welder, chemical plant*

All the company cares about is bucks. They don't care about the people like me who make their bucks. They use people; they kill people. There's nine or 10 people who've died there, and they didn't give one rat about it. All they care about is bucks.

# PART 3

# Coping

# 5    Protection on the Job

In 1979 several fire fighters in Texas died when their NIOSH-certified respirators failed to perform. Three hundred thousand respirators were recalled. The incident sparked a debate on the adequacy of personal protective equipment available to chemical workers and the extent to which such equipment is maintained and actually used. It also fueled a dispute over the relative costs and benefits of personal protection devices and engineering controls.

There are three modes of protection against chemical hazards in the workplace, often used in combination. Sometimes it is possible to eliminate hazards at the source by using fewer toxic substances, using chemicals in the form of pellets instead of powders, or changing the process of production to avoid the manual handling of materials. Where such changes are not possible, harm can be minimized by engineering controls: hoods and ducts, fans and blowers, and generally improved ventilation techniques.

More commonly, exposure is controlled by personal protective devices: gloves, boots, goggles, special protective garments to prevent physical contact, and respirators. Several types of respirators are in common use: vapor filters, dust masks, or alternative air supplies.[1]

1. The effectiveness of personal protective equipment is uncertain. NIOSH tests of so-called impervious gloves found them easily penetrated by benzene and other chemicals. Respirators offer adequate protection only if appropriate to specific hazards, properly fitted to each individual, reliably maintained, and, above all, regularly used by the worker on the job.

Disputes occur when there is pressure to minimize exposure. The choice of appropriate protection follows from the attribution of blame. Organized labor, placing responsibility on management, advocates cleaning up the workplace to eliminate hazards, and providing adequate engineering controls. For employers, the protection of workers against health hazards is an economic issue: the capital cost of retrofitting plants, combined with a view of worker responsibility for health, often biases industry toward personal protective equipment. The question becomes whether to adapt the workplace to the worker or the worker to the job.

The dispute over precautions extends beyond the question of immediate cost. Personal protective equipment places responsibility for protecting health on the workers themselves. Ill health can then be blamed on their failure to comply. Conversely, engineering controls place responsibility on management, shifting both the burden and the blame. To insist on personal precautions is to reinforce the belief that individuals are responsible for their own health and safety. To accept engineering controls is to accept the notion of corporate responsibility. As an issue of responsibility and control, the means of protection remains a source of dispute.

In this chapter, the workers express their views about the relative desirability of various protective measures and the problems they face in trying to protect themselves.

**Taking Precautions**

In some plants, workers are required to wear hard hats, safety shoes, hearing protectors, safety glasses, respirators, or gloves. The manager of one research laboratory actually fires workers who fail to wear safety equipment. However, other people work in situations where supervisors are lax about enforcement, especially if the safety equipment interferes with production. Where personal protection is optional, workers must be able to identify what is appropriate for the job and request the proper gear if they want to protect themselves. Our respondents' comments convey the wide range of behavior with respect to the use of protective equipment.

---

*Stuart, mold maker, glass factory*

About a year and a half ago, after a guy in the shop got silicosis, I started wearing a dust mask. When I worked the bench, my lungs hurt, and after this guy got sick I realized, "This has got to be it. I'm breathing the stuff, I'm blowing black out of my nose. I can taste it in my mouth. I'll wear the mask." It's uncomfortable in the summer heat because you're sweating, but I say to myself, "For the little bit of discomfort I'm feeling here, I figure I'm doing myself a favor.''

   I had one incident. I was grinding on a big bench grinder and the stuff got real hot. When I got all done I didn't feel well. Later I went over and sat down in a chair and felt like a diver with the bends—my knees and elbows felt like there were bubbles of gas in them. It was very, very strange. The guy I was working with came over and asked me what was wrong. I said, "I don't know, maybe I got a whiff of paint or something off that grinder." That's about the time I started wearing a mask. I only wear it when I'm working at a bench—the lathe is not as dusty. When I wear it I don't get the blackness out of my nose when I blow. There may come a time when I have to wear it running a lathe. There may come a time when I should quit smoking. I've got emphysema already. As I'm getting older I think more about it than I did 10 years ago.

### Nick, chemical operator, chemical plant

   In the beginning I was not as conscientious as I am now. Fifteen years at this job has made me kind of leery. What made me more sensitive than anything else was a guy who died in our own local. He was fine one day and then went down like a ton of bricks. They found high concentrations of different things in his system. He worked with some bad stuff including benzene and toluol.

   Anything can be handled if you're careful, but some people just don't seem to realize how dangerous this stuff can be. They figure, "Aw, a few minutes of discomfort, of inhaling something, and that's it." But it ain't just a few minutes. That stuff is in your system and eventually it's going to do something. Your kidneys aren't going to function or your lungs aren't going to work. Most people tend to be careful, but you've got to be continuously reminding them. Of course, you learn more and more as you go down the line. When I first went there we used to run anything through the red rubber hoses that are intended just for water. Now we use chemical hoses. But a lot of precautions are after the fact.

### Les, furniture restorer, self-employed

   I'm now aware that certain things I use are carcinogenic, things that I was totally unaware of before I came to this country. I could appreciate that if you breathe something nasty it could make you ill, but it was a shock to find that certain things I work with, like toluene, can penetrate the skin. So now, I have neoprene gloves, an extractor fan, respirators, and masks. We never used to wear masks. I never wore anything to cover my eyes before I came here two years ago, but now even when I'm grinding a chisel or buffing brass I wear eye guards, just in case a bit of buffing flips off into my eye. I have the extractor fan on when I do stripping. I wear gloves when I'm using a lacquer thinner. For actual polishing I don't wear gloves because it makes the work too difficult. I even find wearing a mask and goggles restrictive. They definitely slow you down.

### Mark, physicist, university

We work in a machine shop, and a guy who I work with quite a bit is sitting there. His clothes, of course, are just coated with little metal chips. Only after his wife rolled over in bed on the metal scraps and got pretty mad about it, did he start wearing an apron at work.

### Mel, silk-screen printer, toy factory

No matter how neat you are sooner or later you're going to get a big whiff of thinner right under your nose. A lot of times when I use it, I just hold my breath, turn around and take a deep breath and hold it again. I really get dizzy doing that. There are times when I just have to stop what I'm doing and get my senses back.

### Ben, repairman, chemical plant

I worry about bringing stuff home on my clothes. I didn't think about it three years ago, but recently I started telling my wife, "When you wash my work clothes in the washing machine, make sure you run the machine through empty the next time." About four or five times the clothes were so bad I told her to take them to the laundromat. There's even been times when I've thrown clothes out instead of bringing them home and washing them. Working with the calcium chloride, I get slimy, and when I dry, it turns to powder. I've taken two pairs of pants like that and thrown them out.

### Earl, landscape supervisor, botanical garden

I can remember the old days when I first started spraying. We used the same chemicals we use now but we didn't protect ourselves at all; no respirator, no head cover, no eye cover, no body cover except the gardening clothes. We sometimes had the good sense to wash our hands. We knew it was somehow dangerous and we ought to be careful handling it. A big change came with FIFRA (Federal Insecticide, Fungicide and Rodenticide Act) when the government said, "This is the way it will be." That really raised the level of knowledge and awareness. The government provided money for training programs and required sprayers to get a license. Now there are seminars and workshops for people working in pesticides; not just to learn about new chemicals but how to keep safe. The awareness of risks does make us more careful. I'm sure that we are more careful than Harry Homeowner who, maybe twice a year, buys a couple of jars of "Spray This" or "Spray That" weed killer. He probably inhales more chemicals in his home than me or my gardeners because we're dressed up like spacemen to protect ourselves. I see my neighbors in their yards just spraying like crazy right in their own drift. They've got to be inhaling some of the fumes and mist. If I spray my own garden, I protect myself completely. So perhaps being a worker in the field really makes you more aware of the hazards.

### Elise, laboratory technician, research institute

Any time we worked with a chemical that had written on it "poison, cancer," we'd do it in the special hoods with plastic gloves and all of that sort of junk. We didn't work with these chemicals routinely, so we were very careful. We weren't as careful with the things we worked with on a day-to-day basis. When you work with them and nothing happens, you become careless. I was never really sloppy, but I was never very cautious. If I got a little tritiated thymidine on my hand I didn't bother to get up and wash it. I just always washed my hands before I went to lunch anyway and wouldn't bother to do more than that. We worked with $P_{32}$ which is a gamma emitter and that always made me real nervous. I always put the tray in the hood and then I had this big plastic shield about an inch think, which my boss had labeled "sterility guard." I would put that in front of me so that my body was protected and the only things that dealt with the $P_{32}$ was my hands. The reason we were so careful was that we didn't work with it much. In the other lab they used a lot of $P_{32}$ and treated it with no more respect than I treated my tritiated thymidine. It's clear that habit makes the mind less wary.

### Irv, plastics fabricator, aircraft factory

I would love to see a more conscientious work force when it comes to using safety precautions. We have people with seniority who have literally forgotten about the hazards of their job. They've done it so long and so repetitiously that they have either forgotten or they have gotten into a pattern where they haven't got hurt before so they do it the same way over and over again. They aren't really aware of where they're putting their hands, or of how close they are to a solvent, or if they're putting the top back on a can that could explode.

### Lisa, laboratory technician, university

At first I tried very hard to think about what I was doing, and about the risks. You couldn't really see anything even if you did spill something, so you had to constantly be asking yourself: am I contaminated, or am I gonna put down the pipette on an absorbent paper, or am I putting things that I touch in the right place? You have to constantly be aware. After a while, my mind turned off. The job got very boring so my mind would start wandering while I did these procedures, and I just kind of let it go. You get inured to the risk. You know how on the road you kind of cut in too close in front of people? Then you see a big wreck and you get cautious for a while.

*Jill, dialysis technician, health clinic*

Well, I take whatever precautions I know about, but you can only do the best you can, and trust the Lord for the rest.

---

### Personal Protective Equipment

Most people recognize the need to wear protective clothing and respirators, but nearly everyone complains about the problems they cause. Discomfort is a major constraint: "I feel like I'm working in a clamshell filled with marshmallows." Respirators can be cumbersome, uncomfortable, and especially awkward for people with glasses, beards, or heads that do not match standard sizes. There are also practical constraints. Gloves interfere where fine work is required. Respirators scare away clients or customers. Bulky protective garments slow down work. Workers may face social constraints from the sarcasm of their supervisors or the teasing of their peers. Some problems arise because the equipment is wrong for the job.

Personal discomfort and social constraints lead some workers to reject available equipment, while others protect themselves. Experience and observation lead some to ignore or deny problems, while others become more cautious. We found no clear patterns that would explain these differences except that those who are better informed about hazards are more likely to seek protection. But even when aware, people sometimes simply make do for the moment: "I just hold my breath." And despite concern, the day-to-day routine of a job sometimes leads to carelessness: "My mind just turns off." For many people, the need to use personal protective equipment has become another source of alienation and isolation.

---

### Discomfort: "Like Working in a Clamshell"

*Bob, firefighter*

With the self-contained breathing apparatus, we can go into hostile environments and have a lot more protection than we used to have. A lot of developments that they made in NASA were passed onto the fire service, like the breathing apparatus and NOMEX cloth. The mask and tank I wear weigh about 40 pounds, and it's only good for 40 minutes. That's 40 minutes without really doing any heavy labor. The more you work the harder you breathe and the shorter time the tank's going to last. Our problem is a tendency to take the masks off too early. If the fire's knocked down, and you can see pretty good, you want to take the mask off. You feel confined, since it weighs 40 pounds and you've got it slung on your back. You're fighting the heat, and it's extremely hard to breathe through the mask because it's so hot. The closest thing I can liken it to is exercising in a sauna. The air feels real heavy like you can't get your breath. The heat's intense, your face is sweating, and you can't see because the sweat's running in your eyes. You're running into walls, and it gets frustrating. You feel like you can't get enough air, and you may really be running out of air and need to change the tank. So if it seems all clear, rather than changing the cylinder you just take the mask off. But with all these plastics and artificial products and chemicals, there may be toxic chemicals still around that you can't see.

### Rich, orchard worker

To be properly suited up to apply the pesticides, you have a full suit of clothes, rubber boots on top of that, then a rubber rainsuit, jacket and pants, rubber gloves, a respirator, a hat, and goggles. When the temperature starts getting up near 80, it is almost impossible. It's incredibly uncomfortable to wear all that plus a respirator that has to be tightly fitted to your face. Being that uncomfortable makes you sick, and so feeling uncomfortable and ill is a normal part of applying sprays without ever having any reaction to them. You just don't feel right doing it from the start. A lot of times it's really easy to drop a respirator down, catch your breath, or pull your gloves off. You shouldn't do that because your hands get exposed and you contaminate the gloves when you put them back on. But you just have to get them off because your hands are so loaded with sweat.

### Earl, landscape supervisor, botanical garden

The stuff we wear now is uncomfortable to work in. When you sweat inside the goggles the visibility is poor. The suits, which are impermeable, are suffocating. It's very hard to breathe through a respirator. You have to have good lungs. We have a yearly lung capacity test for all our workers who use chemicals. We have a couple of people with poor lung capacity, and we don't let them spray because just the effort of drawing the air in and out of the respirators is difficult. No, the gear is not fun.

*Laura, filter cleaner, pharmaceutical plant*
     The masks are terrible. It's a terrible thing to do to human beings, to put a mask on their face. The surgical type masks are the most comfortable, but they're also the least effective. I don't like the 3M mask because you have to pinch it on your nose. If you have glasses and it has any air leaks you fog them up with your breathing. I have a problem with masks because I have a small face. They slide off the back of my head and squish my head up, and there is a big gap under my chin. I told the safety manager that I could put my finger right up under my mask and he told me to wear a respirator, stupid bastard. . . . Now I'm gonna have that rubber thing on my face, which is worse. I get big dent marks, and it makes me break out. I could grow a bigger face.

*Sheila, laboratory technician, research institute*
     I wore a mask when I worked directly with toxics, but the rest of the time, even though I knew they were in the air, I felt it wasn't bad enough. It's so uncomfortable to wear this thing. After hours of wearing it, I'd get these really tight indentations in my face. It was totally inconvenient, I couldn't communicate, I was wearing goggles, respirator, lab coat, surgical gloves, because that's what we were supposed to do. But, how long can you wear that without feeling like you're inside of a clam shell filled with marshmallows?

## Practical Constraints: "It Gets in the Way"

*Eric, sculptor, self-employed*
     I feel bad whenever I use resin. When I repair a piece of sculpture I mix one or two batches. I usually open the window, but I'm still breathing a lot of it in. Sometimes I wear a mask, sometimes I don't, I just figure I'll do it fast enough and get out. Sometimes I wear gloves, but with the tiny patching it's awkward. I'm really not 100 percent careful. When I worked on casting a life-size figure with a big mold, I wore an apron, a work shirt, gloves, and a respirator. I worked in a spray booth with big fans sucking out the air, so I got no fumes. That was a big project, so I could take the time for these precautions. But if I need to do a little job, it would take too much time to set up, clean up, and put the stuff away, so I don't really bother. When you see so many people not being so careful, you tend to say, "Well it can't be that bad." I know it's not true, but I use it as an excuse. It's like smoking. I'll smoke cigarettes every now and then and I'll read the reports and stop, but then I'll start again, thinking: "Well, it can't be that bad, all these people smoke."

### Carol, laboratory technician, university

To prepare for experiments, I have to embed the tissue in paraffin so I can slice them as thin as 20 or 30 microns. The knife has to be very, very clean, and I have to continually clean it with xylene or the paraffin would stick to it. I tried other things that weren't toxic, like alcohol, and they just didn't do the job. I got bad sections. I tried wearing gloves and I just found my work wasn't as good, where if I just used my bare hands I got good sections.

### Debbie, hair stylist, beauty salon

You know, people have been putting dyes, bleaches, and all sorts of stuff on their heads for ages. But the customer isn't breathing it, we're the ones who are breathing it everyday. I never thought of it, really, until I got the lung problem. Well, for the first two years after I got sick, I wore a mask. But it turned customers off so bad. I mean, how can you get a perm put on your head when the person who's doing your hair won't even breathe the stuff? It's just not good for business. They also tell you to wear rubber gloves, which is totally impossible. To give a perm, to make the hair smooth, you have to be able to feel how the chemical is reacting on the hair so it doesn't get mushy.

### Henry, rosarian, botanical garden

Sometimes my supervisor would have me protect myself more than I want to be protected. Case in point—when I'm spraying, he wants me to wear goggles, but I feel that they're really a big nuisance, since I spray down, not up. He also wants me to wear gloves since I go back into the garden area to deadhead the roses right after I spray them. He's right. The spray is fresh even though it's dry on the plants, so I ought to be wearing gloves. But I get a better grip on whatever I'm using if I'm not wearing gloves. So I've compromised. The hand that I handle the rose with is gloved and the other hand with the pruners isn't. It depends on the situation; if the stuff will give me a horrible case of blisters I'll wear gloves, but outside of that I feel like I spend more energy trying to grip whatever I'm working with when I have a pair of gloves on than if I use my bare hands.

### Dorothy, deckhand

You have to rub white lead mixed with tallow into the standing rigging of some sailboats. You'll be hauled up to the top of the mast in the bosun's chair and then you'll run down the rigging in the chair. People will let you off the line as you go down, and so you grip the cable with your hands. The friction heats up the lead and tallow so it melts down into the rigging and into your hands. You're in the bosun's chair with the stuff falling all over you. If you wear gloves, after five minutes the gloves are soaked through. Nobody wears rubber gloves because you can't hang onto the cable. So sometimes people wear cloth gloves, but you can't keep your grip on the ropes. They slip out of the hands.

## Social Constraints: "The Man from Mars"

### Carol, laboratory technician, university

I was ridiculed by a professor from another lab who saw me working with a highly carcinogenic material. It was a benzene derivative, so I wore a filter mask, gloves, and a lab coat. I bought these diapers to spread out on the work surface so that if I spilled something I could wrap it up real quick and put it into a plastic bag. Then the Safety people could carry it off and do whatever it is they do with carcinogenic materials. Anyhow, a professor who had just been hired walked by the room one day and saw me getting all this stuff ready. He laughed and said, "They don't take all those precautions at Harvard, hah, hah, hah."

### Rich, orchard worker

My supervisor's attitude is, "If you feel you have to use these things, if you have to get your respirator, then fine, go get it. I'm not going to tell you no, but personally, it doesn't bother me. If you're just a little careful, you don't have to worry." It gets to a point where you feel somehow you're a little more cowardly than the next guy to go through this whole routine. If the attitude from the top down was one of safety consciousness, then a little caution would prevail. But the opposite is true, so people are scared but try not to show it.

### James, computer assembler, manufacturing plant

One woman, who was a temporary, worked with epoxy and asbestos. She was starting to cough a lot so she asked for a mask, but the manager wouldn't give it to her because he claimed that it was within specs, within legal limits. She offered to bring one of her own in, and he still wouldn't let her wear it because he didn't want it to look like there was a problem. If you see people's happy smiling faces without respirators, well obviously there's no problem. But if you see people wearing respirators, then why did they put those respirators on? It brings to mind images they don't want you to see.

### Kitty, industrial painter, university

I get teased unmercifully because I sometimes wear a mask. "The masked marvel" and all that crap. I wear it when I sand, but I'm the only one who does. Sanding dries you out, and if I'm sneezing and blowing the stuff out my nose, God only knows how much is going down. I showed up at a job one Monday or Tuesday morning and the first thing they did was put me to sanding ceilings and it was falling right down in my face. I asked for a mask and they started laughing at me, "Jesus Christ, she's only here five minutes and already she wants a goddamned mask." They don't like them because they have a funny smell. Also most of the guys on the job smoke and of course you can't smoke with a mask. Then, of course, it's the big macho thing. A guy came down with the flu. He was sick for two days and kept working right through. The third day he fainted dead away on the scaffold. It took that much to make him go home.

### Gene, pipe fitter, chemical plant

I'm very meticulous about what I put on my face. If you had an air mask on and set it down I wouldn't put it on until I'd thoroughly scrubbed it. You could be my wife and I still wouldn't put it on. Being through gas warfare in the service, it's a habit. One time, I took this canister mask and really scrubbed it all up clean, screwed the hose back on and tightened it. When you breathe, the whole mask should cave in, then you've got a good seal. But this time, when I let go, everything went completely black and I dropped right to my knees. I took it apart and saw this black powder, Darco, a filtering agent. Some idiot had put two-three spoonfuls of Darco in the hose and put it back on, thinking that was funny.

## Wrong Equipment for the Job

### James, computer assembler, manufacturing plant

We used to bring in these old computers to a washroom. After we got done tearing them down, we would wash them with a highly alkaline solution. People had to wear protective gear from head to toe. However, we used face shields but no respirators, so the fumes from the alkaline solution would literally burn your lungs. It was really bad. It would get under that mask and nearly choke you to death.

### Kitty, industrial painter, university

The only mask we get is the little paper ones with rubber bands around them. That's all we ever get no matter what we're doing. The paper masks really don't do that much good. You take them off and you see dust on the inside. I was spraying radiators, with Rustoleum, an aluminum paint that contains toluol, and of course when you spray it on a hot radiator it vaporizes. The foreman on the job told me that I should use a mask and he gave me the damned little paper thing. I said, "There's no point to wearing that. That's a particulate mask, it's not going to do anything for the vapors." He couldn't come up with one with filters, so he said to wear it anyway. I did wear it for a while, but gave up on it pretty quick because I knew it wasn't doing any good. When I took it off, it was just as silver on the inside of the mask as it was on the outside.

---

### Engineering Controls

Resenting the need to adapt, many workers want employers to bear responsibility for protecting their health.[2] They want to work in well-engineered environments with ventilation ducts, hoods, and pollution control equipment. But they are also cynical about the effectiveness of such equipment without the organizational changes that would give them some control over maintenance and design.

---

### Fred, chemical operator, chemical plant

When I first worked here, we had to take samples out of the still. So we'd shut off the agitator, and all the fumes would come right up. Guys would cough and gag, with tears and snot running out of their nose. I said, "Good God there's a better way to do this. Breathing all this stuff over a period of time is going to kill somebody." So we all got together and somebody came up with the idea of having an exhaust hose. Different things were suggested, but the exhaust hoses sounded like the best. So we got the company to install these big elephant hoses with an exhaust system in the back and in the front. They have hoses now that run to each still. You take these hoses and set them right up there in the bull's-eye and it sucks out all the vapors.

### Laura, filter cleaner, pharmaceutical plant

There's certain engineering things that I believe they can do. They can afford it. Everybody in upper management should take a ten percent cut in pay if that's what it takes because that wouldn't make or break them either way. Maybe they won't be able to get their new Mercedes this year, but I'd rather see that than see somebody die. They waste so much damn money on a lot of things. Their biggest bitch is wages and the electric bill, but they'll leave every light on in the place and keep

2. These views have had some impact on government policy. For example, during the Carter administration, OSHA's policy on "Identification, Classification, and Regulation of Potential Carcinogens" stated, "Engineering and work practice controls are the only reliably effective means of protecting employees from potential occupational carcinogens." In contrast, the Reagan administration has sought greater emphasis on personal protective equipment, consistent with the belief that workers should bear primary responsibility for safety on the job.

machines running. They take better care of their lab animals than they do us. We're more expendable and cheaper than engineering controls. They expect to make money off of everything they do and they probably wouldn't by cleaning up the shop. But maybe they would. They'd have workers who are in better shape, who are healthier. Maybe they wouldn't be paying premium rates for overtime when people are out sick. People are getting bronchial problems and allergies in the chem sections where they're working with a lot of dust and powder. I think it's an investment they have to make.

### Tony, dry cleaner

To protect yourself, you're supposed to use your head. For example, if you opened a big 50-gallon drum of kerosene and gasoline, you wouldn't stick your head down there and take a deep breath. Some idiots do those things. It used to be that to handle the clothes from the dry-cleaning machine you drew a deep breath, grabbed the clothes, and put them in a recovery place. Then you breathed. Now improvements have been made. Now you don't have to do that. You used to have to transfer clothes from the washing machine to the extractor to a dryer, and those things all let fumes out into the air. Now it's what they call dry to dry. All the things take place in one machine. You open the door at the end of half an hour and it's all done, theoretically clean. Very few chances for fumes to get out, and what few do get out are recovered by a "sniffer." This is a thing that takes the fumes and condenses them into a solvent again.

### Lee, stage carpenter, university

The scenery is all painted in the shop or in the theater, but we really do not have anything that could be regarded as an adequate ventilation system. We have fans in the ceiling which do change the air in the room, but they aren't ducted to any particular areas. The air evacuation is not fast enough to prevent people from breathing in the dust from sanding or the spray from aerosol paints. We really need to have the machines ducted to the fans.

As far as the spray painting, we paint things 22 feet tall and 40 feet wide and you can't duct something like that; you can't do that under a hood. You pretty much need a paint room dedicated to it. But even if everyone considers these things vitally important, we'll wait a long time. It's very difficult to forego, say, adequate seating, when that is what people are paying tickets to sit in, in favor of getting rid of something that no one ever sees, and doesn't have any visible or immediate effect. It's difficult to justify safety to people who actually push the money.

*James, computer assembler, manufacturing plant*

In the area I work in, there's hardly any air conditioning and it's really hot. They could install fans, but they wouldn't spend the money for that. They figure it's cheaper to give us respirators. The building is being rented, so they don't want to put new equipment in. We suffer for their money.

# 6    Adaptations

How do people cope with dangerous jobs? How do they persuade themselves to accept risks and adapt to conditions that they know will affect their health? Behavior in response to workplace risks ranges from carelessness to caution, from denial to protest, from resignation to activism. People's responses reflect their specific experiences at work, their personal economic commitments and constraints, their perceptions of occupational choice, their trust in the system, and their identification with the goals of the organization. In other words, attitudes about risk have as much to do with social conditions, professional commitments, and, above all, feelings of personal control as with the actual degree or potentially catastrophic nature of the risk itself.

The growing literature on risk emphasizes the importance of such factors as the voluntary nature of exposure to risk, the familiarity of the hazards, and the character and extent of the risk itself in determining acceptability. Studies suggest that people tend to underestimate familiar risks, often considering themselves personally immune. Similarly, they are inclined to dismiss the risks of voluntary activities such as skiing or smoking.

Some studies of risk acceptance examine people's preferences as revealed in their everyday decisions and the social situations in which they are involved. These studies assume that existing social arrangements, such as the wages accepted for hazardous jobs, reflect deliberate choices where there are open alternatives. Other studies elicit preferences through surveys. Assuming that individuals act autonomously, these analyses of risk seldom consider the social and institutional context to which people must adapt.

Our interviews complement these approaches by focusing on the social factors that influence attitudes and behavior toward occupational hazards. Most people accept risks because they need a job. While individual personality differences may account for some variation in risk attitudes, pressures at the point of production and possibilities of control clearly help to shape what workers define as dangerous, to constrain their choices, and to give direction to individual responses.

In this chapter, workers explain why they continue to work in environments which they know are hazardous, and how they adapt to situations they cannot control.

### "It's Just a Normal Part of the Job"

Most people go to work each day expecting to do their job and think very little about attendant risks. When they recognize problems they often simply accept them as an inherent and inevitable aspect of work: "It's a normal part of the job." Risks are "a fact of life"; "all jobs are risky"; "it's part and parcel of work." They use language in a manner that normalizes discomforting uncertainties, transforming mysterious respiratory ailments into familiar "chemical colds" or long-term allergies into "rashes." In these ways they make the necessary personal accommodations to conditions they have little chance of changing.

---

### Joe, laboratory assistant, chemical plant

People were getting hurt all the time, so nobody thought it was a big thing for somebody to get hurt, or for a valve to fall out and 1500 gallons of something to come pouring out. That was a normal everyday thing. Nobody thought much about it.

### Dorothy, deckhand

When I talk about working on boats, a lot of people think I'm crazy because they're really into their desk job and only have to worry about chalk dust. But for a lot of people who work in boats, the risk is no greater than the diseases you pick up staying in your hometown. Also, even if it is, you're getting paid. They accept the risk as part of what they're getting paid for. If it's bad weather and somebody has to go forward to fix something that's loose, that's just a part of the job even though you know that there's the risk of getting scraped off the deck by some great green wave. That's a part of the job, and no one ever tried to fool you about that. Some people are really into this macho excitement trip: the risk just makes work a little more interesting. It's the young stupid ones who are into the macho trip, the excitement of the game. Eventually you either wise up or drown!

### Les, furniture restorer, self-employed

As an apprentice we used to work in the stain buckets with naptha-based stains. We got a brown mahogany mark up to our elbows as we dipped into this stuff all day. Some of us used to feel a little bit sick. They'd say, "Just go outside and get a little fresh air," so we'd go out and have a cigarette and come back in when we felt better and off we'd go again. No one complained. What we were interested in was wages. Most people working there wanted the job, and the risks were just accepted as part and parcel. Of course, you've always got the old guys saying, "Well, in my day, it was much worse." The attitude was, any kind of physical reaction that you had was your own fault. If you were to wear a mask they would have considered you a fag. If you looked at a big vat of brown mahogany stain, it was just brown mahogany stain. The fact that it made you feel ill when you breathed didn't occur to me as bad, no more than catching the flu is really bad for you. It's just something that happens.

### Daniel, chemical operator, chemical plant

When I first started working I was scared to death when I saw cyanide, hydrochloric acid, and all that high-powered stuff. But now I'm used to it. I know what's around, what I'm working with. The jitters go away. It don't scare me. If something does happen, I pretty much know what to do: run like hell! You don't want to stand there too long; the cyanide will wipe you out in a matter of seconds. We joke a lot about the risks. We got a lot of rodents that run around, little mice and stuff, and they're always chewing on something. Someone will say, "Hey, watch out for that little rat. He's got cyanide in him and he'll explode."

### Bess, diffusion analyst, manufacturing plant

I think men don't fear supervisors quite as much as women do. I mean, women are brought up to believe what their father says goes, and when they get married, what their husband says goes. Then, in the working world, the foreman is a male and when he says, "That's where you're going to work, and that's what you're going to do," they don't buck it as much as men. If they feel that that job has to be done, then they do it.

*Mike, photo lab processor, blueprint company*

A job's a job's a job. You go there, you do what you gotta do, and then you go home.

*Art, laboratory technician, university*

We'd never get any work done if we had to wait around to find out if things are safe. This is carcinogenic, that stuff grows toes on your feet or something. . . . After a while they'll say you can't breathe air because it's carcinogenic.

*Mark, physicist, university*

Are we supposed to seal ourselves in a glass booth and wait 'til somebody says, "Okay, it's all right to breathe," before we open the door to breathe? I mean, life is something that involves mostly personal judgment. All this safety business is a matter of trying to make guidelines of what to do and what not to do. It's not like I'm dancing through the lab with radioactive chemicals in one hand and God knows what kind of acid in the other—kind of juggling and trying to balance a basketball with one foot while the other guy is clothed in goggles and a hood and 35 pairs of protective clothing. If there's something you have to do which requires the use of certain chemicals, then why should you let it hang over your head? Say you have to commute into work everyday, and you work in New York City. The probability of being involved in an accident is enormous, but you don't call up your life insurance agent every morning before you leave for work. Chemicals are pretty much the same way.

*Irv, plastics fabricator, aircraft company*

I've worked in fiberglass for over 10 years; the resin, the solvents, the cloth, nothing affects me physically. I've never had a sore, I've never had a blister, I've never had nausea from the smell. I don't enjoy it. It's an uncomfortable situation to work in, but it's a fact of life that you just can't get away from the odor. If it's not in your nostrils, it's on your clothes. Even if you wear the company's overalls, you still end up with that odor on your clothes. My wife can tell when I walk into the house where I've been working. She'll look at me and say, "You were in the fiberglass shop today, weren't you?" She can literally smell me coming.

*Ted, welder, chemical plant*

You have to test the tanks before welding them. Before the safety committee was around, we used to tie a torch onto a long pole and hold the torch over the tank to see if it was safe to enter. We don't do that anymore. I was blown out of a tank once, and I'm still here to talk about it. Burned my eyebrows off and screwed up my eardrum. I've got a 40 percent hearing loss in one ear and I burned it too. As my eyebrows grew back, so did my safety awareness.

Still, there are times when you have an "Aw, the hell with it" attitude, where you take your really big risks and don't look out for your own safety. You have to live with it every day, so after a while you're either very fearful or you say, "The hell with it, I've had a good life, let's give it a shot, give it a go." There's been times when I've put a torch over a top of a tank and been extremely afraid, and other times when I've walked over to the tank and tossed a match in and said, "The hell with it," just to see the reaction of the people, especially the boss, to see them wet their pants a little.

---

### "What You Can't Change, You Accept"

We found many people, unable to control the conditions of their work, resigning themselves to accept situations they recognize as grave: "Everyone's going to die sooner or later"; "If this doesn't get you, something else will." Convinced they have no recourse, they feel little incentive to actively seek improved workplace conditions. An extensive literature on powerless groups (poor people in Appalachia, prisoners, colonized people) suggests that such resigned and even complacent responses are common in situations perceived as beyond control. It is simply unreasonable to worry if nothing can be done.

---

*Nora, graphic artist, blueprint company*

I have spells of anxiety same as anybody else, I suppose. I've been exposed to chemicals everyday for over five years now, and I'm sure if it's going to do its number it's going to do its number. I can't go take my lungs and shake them out or anything.

*Fred, chemical operator, chemical plant*

There's a lot of things we can't control, like economics. We've got to work for a living and we're not talented, gifted, or rich. I think that the wealthy are favored in this country by quite a bit. I see things that bother me, but you've got to recognize that you can only try to change what you can change. What you can't, you have to accept. Sometimes we kid about it. "Cyanide is good for your health," "Good for what ails you," stuff like that. After a while you just get used to the situation.

*Ann, silk-screen supervisor, museum*

Sometimes what I do is hold my breath. It's like, how could the fumes possibly go in when I'm shutting myself so tight? Isn't that foolish? I guess, I probably figure, everybody's going to get it one way or another. It doesn't have to be cancer, but it will be death. I'm not ready to quit my job and go through all the bullshit I'd have to go through to get another job when I'm not sure whether it's killing me. It's so easy for me, it's so convenient, and it's only across the street. My field is art and whatever other job I'm gonna get it's gonna be the same. Maybe it won't be silk-screen inks, but if I do paste-ups and layout it's going to be rubber-cement thinner.

*James, computer assembler, manufacturing plant*

Only a few people care about the conditions here. Most tend to think, "It's not going to hurt you," or "What the hell, you could get hit by a car tomorrow," or "Well, if this doesn't get you, something else will." It's a fatalistic attitude. Those who are bothered are scared to death to say anything. In one area a lot of women came down with a rash. I think they were working with a degreaser, probably methyl chloroform. They were all afraid to even bring it up to their manager. They were afraid of being demoted to another job or being labeled as troublemakers.

*Daniel, chemical operator, chemical plant*

I hear stories of people who work in the other section getting cancer from the ketone, but you don't really know. At first it scares the hell out of you. "Wow, the stuff's gonna do me in. What am I doing here?" I guess everybody nowadays will get it sooner or later. I hope I'm not one of them. What can you do? If you're gonna get it, you're gonna get it.

*Frank, chemical operator, chemical plant*

We have accidents with the cyanide. Sometimes it's due to some-one's mistake, like a tank overloaded; other times it's mechanical. When we have to load cyanide, we're always double-checking things before we even start to make sure there's nothing wrong. I try not to worry. It doesn't help the job, and there's nothing you can do. I just try not to think about the future. You know, if it happens, it happens. I have to stay here at least 10 years to get some kind of guarantee of security when I retire. I've already put seven years of my life into the chemical plant. Walking out now would be bad news for me, a waste of time. After 10 years I'll go find something different. But if I stay here until I retire, I'm not going to see very much retirement. I don't think there's more than 10 people getting retirement checks right now. All the rest are dead.

## Stuart, mold maker, glass factory

We've had one guy with silicosis. Do they talk about it a lot? No. Do they seem extremely scared about it? No. One guy went to the doctor who found a couple of spots on his lung. The doctor thinks it's silicosis, but his major worry is high blood pressure. Every once in a while somebody will get going on a health problem and the whole shop will pick up on it real strong. We're constantly bugging them about the dust collection system, but you get tired of beating your head against the wall. Eventually they're going to do something, but when? Yet, it's generally a good place to work. I personally find it boring, but the money's good. There's good times and bad times. It's paid for my house and fed my family. So I sit there and say to myself, "You've been hanging tough for 15 years," and decide to stay for a while more. But then, if it's going to kill me, I've got to think about leaving. That's a tough decision.

## Ken, electrician, chemical factory

Death is a very serious thing to me. I lost a daughter last year, and as I get older, I'm 46, it's something I got to face. I want to live to be 80 or 90 years old, but I might have cancer in me as a time bomb. The only way I can protect my health is to change my life-style. So, I watch my diet, I have my wheat germ and my apple, my orange and my carrot. My carrot is my cancer stopper. They say carrots are good for you. They contain vitamin A. I don't drink coffee or eat cake. I eat my apple and my orange. I walk two miles to work. It sounds stupid, but that's a very unusual thing. People think I lost my car insurance, but I walk by choice. I walk in the rain. The only thing I can do is be in good shape, so maybe my body can fight things off. That's the only control I have.

I can't stop what's happening at the plant. If I had to work in a room that had toluol fumes for three days, I don't think I could refuse if other workers were accepting it. And if slight traces of toluol and benzol give me cancer, what can I do? I believe that mental control and physical shape are the only hope I have against cancer. Because cancer does frighten me, it frightens me a lot.

## Sid, chemical operator, chemical plant

You walk into the lunch room and it sounds like there are a bunch of horses in there. Everybody's coughing. If you ask why, they say, "Well, I work at ——— , that's why I got the cough." If somebody's telling you, "You shouldn't drink that because it's got saccharin in it," they say, "Well, what the hell do I care if it's got saccharin in it? I work at ——— ." There's always jokes about the chemicals. It's something real and people know it. But let's face it. If you really sat down and thought about all the things that the work is doing to you, you'd have to get out. You'd quit. So the right thing to do is get involved, like some people do. But most just get it out of their mind. They just don't think about it. People are more worried about getting burned or inhaling something than about any long-term sickness. Since it's not happening now, it's not something to worry about. You always think, "I ain't gonna be here long. I'm only here for a couple of years." When I first came here I thought five years, that's all I'll be here. I'll be here five years next month, and I have no plans to leave. Most everybody that I talk to says the same thing, but 95 percent of us stay.

### Steve, railroad trackman

Most people aren't that concerned about risks. They're very stoical about these things. There's so many risks that they say, "Well, what's one more thing going to do?" It's as if fate is in somebody else's hands. Railroad workers have known generations of exposure to injury and death. That's part of their life. It's deep in their consciousness to expect the possibility of injury and death and the inevitability of retiring worn-out, on the scrap heap in a semi-crippled state. You just don't take traditions like that and throw them out the window in two minutes. Many workers see their fate as something dictated by God, which I think is very unfortunate. I think the situation is changing, like so many other things—like awareness about relations between races or between sexes. Well, this has to do with a change in awareness about relationships with employers.

---

### "What's Our Alternative?"

The willingness to work in hazardous conditions reflects people's sense of limited alternatives. According to the Michigan survey, many workers feel locked into their jobs. In 1977, 41.9 percent of their respondents felt it would not be easy to find a similar job, and the situation has certainly deteriorated with the increased unemployment of the 1980s. As alternatives recede and jobs get scarce, many workers simply adapt to risks rather than seeking unlikely change.

Workers talk about the risks of their jobs in this context of alternative employment opportunities. Some believe that their training and abilities limits their job choices; others regard the known and familiar risks of the present job as more acceptable than the unknown hazards that could plague them in another plant. Still others, highly paid chemical operators, are reluctant to risk a change that would reduce their wages. Often workers become aware of potential health problems associated with their jobs only after a lengthy period of employment when they have built up time toward their pensions. They cannot afford to jeop-

ardize their future economic condition. Burdened by family obligations in an uncertain economic and employment climate, few
are willing to change their job, or even to risk being fired for
speaking out about workplace conditions.

---

### Arnie, chemical operator, food processing plant

You never balance the wage against the risk; you balance the
wage against the alternative. And the alternative is starving when you're
put in this situation. That's what's so phony about this cost/benefit analysis. A worker in the plant doesn't say, "Well, I'm getting $6.50 an hour
so I'm gonna take this risk." The worker in the plant says, "I'm getting
$6.50 an hour. If I open my mouth I might get nothing an hour, or I might
get minimum wage. In that case, I can't afford to live." So, what's the
difference? There's no difference for a person in that position. Either way
they're trapped.

For example, at the job I had before the food processing plant,
there was a guy who had a perforated septum. He showed it to me and
says, "Look, it's the fuckin' fumes." I asked, "Why don't you quit?"
He says, "Well, I'm 46 years old and you're 18. I've got a third grade
education, three kids . . . where the hell am I gonna go?" I found out he
was right. Because when I quit there and went to the food plant, there
was no fuckin' difference. It's like going from one horrible place to another
horrible place. You're kind of locked.

### Steve, railroad trackman

You accumulate seniority, and that gives you certain prerogatives
and privileges. You accumulate vacation time, you accumulate sick time.
Jobs get harder and harder to find every year. That makes the job more
valuable to you every year. You also get closer to retirement, and that's
a great incentive not to leave. You'd be throwing a lot of working years
down the drain. In our case, we can collect a pension negotiated by
contract as well as railroad retirement benefits in lieu of social security.
If I were to leave now, I would not be able to collect the railroad retirement.

### Joe, laboratory assistant, chemical plant

Going back a few years, I probably would have left, but not now
with the economy all screwed up. People now just can't leave. Where are
they going to go? We have guys who are quitting and going to Florida and
to California. Four months later they're back, begging for their job. It's
the producer's market now. There are people on the streets who would
do anything to make money. It's tough to leave a job and find anything
half decent. I'd rather not have a job than be dead, but because of the

job situation a lot of guys are intimidated, so if you refuse to do something the guy in back of you will go ahead and do it instead.

### Ben, repairman, chemical plant

They asked me to work in the lime kiln and I got stuff hanging all over me, white, drippy, slimy stuff. I got it all over my hands, my clothes, and everything else. Then I said to myself, "Not only am I taking a chance by getting inside the tank, but I'm also taking a chance because I don't know what this stuff is. No, it's not worth my $9 an hour to do that." Most guys won't tell their foreman, "I'm not going to do it," because they just got hired and they'll lose their job. Most of them have been working for $4 an hour and now they're making $9, and they've got good benefits and they're not going to lose that for anything. If you have a family, two kids, a house, a car, and bills to pay, you forget your personal feelings about certain things. Nobody wants to do things that they feel are wrong, but then again they have the responsibility to provide for their family. We really don't have a choice. I can't refuse to work knowing that tomorrow I can get another job. I can't look for a year and a half for a job. I'd lose everything. I don't think it's right for them to say, "Well, if you don't like it, leave."

### Sandy, rigger, chemical plant

Every worker has a choice. Any worker can quit his job. I mean, when you come down to brass tacks, anyone can quit. But the realities of life—family, the children, mortgage payment—impose certain limitations on a worker's right to just quit. I don't feel personally that people should have to quit to protect their health. I feel that the employer by obligation, by law, must provide a safe and healthy workplace. And if the employers live up to their obligations, then there would be no reason for a worker to make the choice.

### Sheila, laboratory technician, research institute

One woman who complained the loudest about some of the safety problems got pregnant and worked her full term. I tried to talk to her about it and she said, "Yeah, I know, I know." "Well, I don't think it's that bad, really . . . ," and that was it. She really didn't want to talk about it. Her attitude changed because she needed the money and the only way to carry through with this idea of having a child was to have lots of money. So she just put on the blinders and worked.

### Don, railroad conductor

If I had known that it would be that lethal, that it could give me or one of my children cancer, I would have refused to work. But it's a matter of survival and we just can't consider all these things. Of course, now I'm not going to smoke. But I'm probably as careless as anybody else. You would think after going through something like a daughter with cancer, that I would have common sense. But I still buy that red stuff in a can that you mix with water, Red Dye No. 2, or whatever it is. I still eat bacon. I did get radical at one point, though. I wouldn't eat at anyone's house. I actually got the taste of sugar out of my system. But the minute you walk out of your house it's shot. The minute you start your car up, you're breathing those fumes. We shouldn't buy plastic bags. It makes me angry when I think about it because we're trapped. Something's going to get us anyway. Meanwhile, we've got to make money to survive. We even fear that they're going to get us, that we'll be fired if we open our mouth. If we make waves, it could be bad. I need the job, I have a mortgage and a lot of bills. So I put it out of my mind. It's not until it hits you, right between the eyes, that you say, "My God, it's too late."

### Mary, housewife, wife of a railroad conductor

I saw a show on television, a documentary about job-related incidents where people were working around carcinogenic substances. There was a man, 50 years old, who had worked in an asbestos factory and was dying of lung cancer. He was in the hospital dying, and they asked him, "How do you feel about the fact that this was caused by asbestos?" He said, "Very bad." Then they asked, "What would you have done 20 or 30 years ago when you first got this job, had you known that it was going to lead to this?" He broke down and cried. He obviously had a lot to think about. When he got his composure back he said, "I don't know." That was the end of the show. He honestly did not know what he would have done if he knew how he was going to wind up.

At that time, before the baby died, I thought, "How absurd. Are we that trapped that we would take anything just to have a job?" But then, if we didn't have a job, where would we be? We make much more money than a lot of people we know with a lot of education. How desperately we need our jobs in this country! We're really trapped. What would we do if we said, "Well, because of the dioxin, we're not going to work for the railroad." Where would we go? Where could we make enough money to support a house and all the other bills that we have? Where are the options? There are no options. So we keep our mouth shut. We never talk about the issue with our friends who work for the railroad. We don't try to keep it quiet, I don't mean that. It's just nothing to talk about. Who would listen? And when we do mention it, they say, "Oh, you sound like those Vietnam veterans," or, "Shut up, we don't want to hear it." That's the attitude. "Worry about what you're going to do in the future"; "Have another kid"; "Don't worry about what caused it"; "You're never going to find out for sure."

### Mike, photo lab processor, blueprint company

You know, the guys who work with me, they have families. One of them is very concerned about what we're working with, but he has financial commitments. He just wants to make it every month, or somebody comes banging on his door. So he can't say much. I'm not married and I don't have a family. I have a lot less to lose. So if something's wrong, why not say it?

### John, maintenance worker, food processing plant

In a way, I'm kind of a loose bolt in the organization. Managers fear me, because I'm wildly overeducated for the job. I'm not the person to mess with. I'm a very mouthy guy and try to keep up with the facts, plus I can just quit anytime I want and go somewhere else. I have no responsibilities, no payments, nothing.

---

### "It Won't Happen to Me"

Workers who cannot accept the idea that their livelihood could also be a source of harm cope with their situation by denial. They build fantasies about their invulnerability: "It won't ever hit me." They joke to reduce tensions and to block their fears. Some deny dangers simply by refusing to think about them. Faced with unavoidable hazards that they cannot control, they essentially deny them. We found denial, even more than other adaptations to risk, mainly among people who had no expectation of either controlling their working conditions or of finding other jobs. But on occasion, for example the party in the room of dioxin samples, denial simply represents thoughtless behavior.

---

### Sandy, rigger, chemical plant

People joke about the stuff they work with because they're not willing to face the realities that they are harmful. They try to pad the situation by joking about it. They seem to have this theory that they're protected by some invisible force field, an individual force field that will protect them against occupational disease.

### Dorothy, deckhand

I think the main reason why people on ships don't worry about risks is that they don't know any better. Getting rust all over you and breathing in chemicals is just like handling dirt, right? So, did you worry about dirt when you were a little kid? Of course not. They worry more about lines breaking and stuff like that. I joined a ship where a guy had been decapitated when a winch cable broke. They were still talking about that. Those kind of injuries are much more striking, much more obvious. People get rashes all the time and who knows what it's from? You just ignore it.

### Penny, laboratory technician, health department

At the toxicology lab we were testing samples of dioxin-contaminated soil. We kept all the samples in a large vault. One time we had a Christmas party and because this vault was soundproof and we could lock it from the inside we held the party there so we wouldn't get caught. They brought in food and drinks and actually made screwdrivers in these big separatory funnels which we normally used to separate chemicals. They had food—ham and cheese and dips and chips—sitting on boxes of the dioxin samples. All the lab technicians, including the supervisor, had themselves locked in this room and had themselves a grand old party. Actually when I think about it, it was bizarre. The whole time I was in there, which wasn't really that long, I was feeling like this is bizarre. They had thousands and thousands of dioxin samples. The room was jam-packed full.

### Stuart, mold maker, glass factory

The easy way out for me is to wear a mask. I get kidded about that—"Hey, doctor!" Here I am with my hands looking like I've just come out of a coal mine, with my mask on. But most people don't even bother to wear masks. I've passed the right-to-know stuff around the shop. Two or three guys looked at it and threw it down. They say, "Hey, you're scaring me. I can't come to work anymore." They don't want to accept the idea that coming to work could kill them. Yet sometimes they talk. "He just retired and he's dead," or, "He was only 56," or, "Does everyone who gets out of here die early?"

### Art, laboratory technician, university

For some reason, all my life I have had to prove myself that I wasn't a coward, probably because I think I'm more fearful than anyone else. I walked across a suspension bridge rail when I was a college student. It was a small suspension bridge, and the rail was real narrow. I was on a bender, and you think, it can't happen to you. When I was in the army, I was in artillery, and I heard that there was a bomb disposal job open, so I got it. First I felt that I had been able to prove that I wasn't a coward, but later I realized I was more of a coward than ever. There was the time, years ago, when I thought, "Gee, I'm doing all these hazardous things all the time; I ought to get hazard pay." I was feeling a bit resentful because I had to take chances. Now, as an organic chemist I don't think twice about going in and doing something hazardous. Every new reaction you run can blow up in your face for some dumb reason. It's part of the game. I'm sure it's better to be safe than sorry. I understand that, but I'm not sure how that translates to human beings. There are people who work with carcinogenic agents all their lives and nothing's ever happened. There are people who smoke and get lung cancer and other people who don't. Then, there are a few who don't smoke at all and end up with lung cancer anyhow. The question is, How do I relate to all this? What's going on?

*Gene, pipe fitter, chemical plant*

Over a course of a month's time, we must be exposed to a hundred different chemicals ranging from caustics to cyanide. You don't really think about the dangers. If you really sat down and thought about what you're exposing yourself to and what it can do to you, I don't think anyone would work there. But the human mind is strange. It blocks stuff like that out and you don't really think about it. You think more about the petty aggravations, the stupidity that you have to put up with. That type of thing.

*Eve, sorter, manufacturing plant*

I had a girl working right next to me on a big machine where there were different acids, and she was a very sloppy worker. I told her many times to wear her lab coat and safety glasses and to quit being so sloppy. When she had to use our sink to wash out acid bottles she would leave acid on the sink, and then when I used the sink I got burned. Some people are their own worst enemies. They are told by supervisors and the union how to use things, but they just don't think anything is going to happen to them.

*Sheila, laboratory technician, research institute*

We worked with things like methyl mercury that you can't smell or see but that have a devastating effect on the central nervous system. At first the technicians got scared when they found out. Everybody got into a panic and wanted something done. They would come to me and ask me to do something, and I would go and scream to the appropriate people, and nothing would get done. But, after only a few days the concern seemed to dissipate. While people know there are long latency periods on a lot of these exposures, it was like, "Oh, gee, it's been a few days now and nothing's happened to us."

---

**"It's Worth the Risk"**
A number of people, mainly those in professional and skilled jobs, told us frankly that their work was "worth the risks." Aware of the hazards, they accept them as a trade-off for the personally gratifying benefits of their jobs. While often very careful to protect themselves, they measure the risks against the satisfaction of their work and the priorities of their careers. Fire fighters feel the risks are small compared to the satisfaction of saving lives. A deckhand is willing to take risks because she values her autonomy. Artists value the opportunities for creativity. A painter and a rose gardener love the aesthetic quality of their jobs.

---

*Eric, sculptor, self-employed*

I'm aware of the problems, but what I want to do is make sculpture, I don't want to find the safest way to live. Otherwise I wouldn't be in this city. Am I compromising my health for a buck? No, not really. What I do is more a labor of love than money. Getting the thing done and getting it done right is of prime importance. So I try to compromise. For example, I'd like to work with less resin and do less grinding on seams. But the caster who is good enough so I don't have to grind the seams is much more expensive than the guy who does it now. If I had the money I would send it to the more expensive guy because I don't ever want to smell resin again. But in a sense by working on the cast myself, by grinding the seams, I get very intimate with the final sculpture and I make changes. I'll grind the eye open a little bit more or turn the mouth a little bit more. In a sense by being forced to work the resin, it makes a better sculpture. So I don't think I'll ever be able to get away from it.

*Mark, physicist, university*

I work in low-temperature physics. We study properties of materials that are very cold, which in our case would be below 1 degree Kelvin or about 450 degrees below zero farenheit. It's hard to call this a job. In any kind of experimental work, the whole idea is you are not doing it because it's a job, you're doing it because it's something you're interested in. You're working on a project, and whatever comes, you do.

*Bob, fire fighter*

Fire fighting's the most dangerous occupation in the world. I've met fire fighters at international conventions from Hawaii to Ottawa, from California to New York, and they're a special breed, there's no doubt about it. Maybe it's macho. Some of it, I suppose, is because you're pitting your skills against one of man's greatest enemies. Rescue is part of it. We save people's lives. I've done it myself. That's a hell of a feeling, something that nobody can ever take away from you. You've done it all by yourself. If it wasn't for you, this person would no longer be a member of the human race. There's not too many people who can say that, outside of doctors. It's worth the risks because you know you are doing a worthy service. I've seen fire fighters killed, and when that happens I say, "No, it's not worth it." But they were just like me. If somebody would have told them, "Come on now, you're taking a lot of chances," I'm sure none of them would have left. So, like a psychologist would say, "If you're happy with your work, you've got it made."

### Bill, fire fighter

The risks are always in the back of everybody's mind. If they weren't we wouldn't be alive. At the same time, they're put in the background because of the nature of the job. The primary job is rescue, saving people's lives. The secondary job is saving what these people have worked for most of their lives, their home, their property. That outweighs the hazards. It really does. The hazards are there, but they're not as important as some little four-year-old who's off in a closet someplace scared to death because he set his house on fire and now he doesn't know what to do. I guess if there's a chance that you could save somebody's life, your own life kind of gets put in the background.

Not everybody can be a brain surgeon. I think that there is a lot of self-gratification in doing the job and doing it well. There's a lot of camaraderie, too. You're dealing with men who work good as a team. When they have a problem, they want to solve it, and they do it professionally. I enjoy that part of it—seeing that type of team work and being a part of that group, sharing the satisfaction of knowing you've done a job well.

### Kitty, industrial painter, university

I love my job. I'm the kid who always used to make Dad slow the car down when we passed road crews. I was always fascinated with buildings and construction and especially painting. It's dirty but it's fun, almost like a second childhood. You get dirty and make noise all day, cuss and swear. A lot of people tell me it probably ought to be women's work anyway, because it's making things pretty as well as protecting them. And it pays good. People don't realize how valuable painters are. They think we're just cosmetic, but it's much more than that. Buildings would fall apart and you'd never have a decent looking piece of woodwork if the painter didn't varnish it.

### Dorothy, deckhand

In any job, there's certain things you have to put up with. I'd rather stand on an icy deck than be a waitress and smile at businessmen who say, "Hey, sweetie, come over here." It's a question of what sort of things make you uncomfortable and what you're willing to tolerate. Some people would rather accept the risk than spend their whole life walking behind the rear end of a donkey with a plow. If you're worried about the quality of your life as well as the length of it, you may be willing to incur a calculated risk.

### Irv, plastics fabricator, aircraft factory

I'd like a nice, clean, odorless, dustless area to work in like everybody else, but the work itself is enjoyable to me. I work with my hands everyday. I'm not much of a desk person. I feel that I'm a skilled workman because technically fiberglass is skilled labor. So I'm satisfied.

### Henry, rosarian, botanical garden

I love what I do. Probably would do it even if I wasn't getting paid for it. So I don't really equate the pay with the risks that I take. I don't really feel that I'm taking that many risks. I just don't feel like I'll be plagued with problems from working with pesticides. It's just my religious belief.

Other people work for a paycheck. I don't really picture myself as doing that. I work more for the beauty of the gardens. I enjoy doing this type of work, making things look good so that other people can enjoy them. It's a combination of working with plants and talking with people at the same time. Whenever people ask me how long I'll stay here, I tell them that I'll stay here 'til something better comes along. Haven't found anything better yet.

### Les, furniture restorer, self-employed

The furniture I work with is stuff that I could never hope to own. But I love the stuff, I love to look at it and play with it. I get a buzz out of doing a nice job on a nice piece of furniture. It gives me a lot of satisfaction. In many ways I'd go so far as to say that I'd rather die 10 years before my time and have had some job satisfaction than work at GM and be bored to death. I don't actively wish to use things that are bad for me. I've looked around at possible alternatives, but they don't produce the job. I got some work in the museum, and it's in there because I've used materials which are, if you like, carcinogenic; there's no way I could have done those jobs using totally safe materials. So I could say if I really went out on a limb that it's worth the cancer problem to conserve pieces of historical importance.

# 7    Activism

A growing number of workers are becoming health and safety activists, seeking ways to improve the conditions of work. Some of these activists serve as health and safety representatives of their unions. Others collect information on their own and serve informally as resources for co-workers who are concerned about particular health problems. Most (but not all) of the rank and file activists we interviewed work in unionized plants where they feel some degree of protection when they take action on a health and safety problem.

Workplace activism has a long, though largely unexplored, history. The original union activists were organizers struggling to form new unions in the face of powerful resistance. While "organizing the unorganized" remains a current slogan, most union activists today operate in a different context. They are members of an existing union with its own bureaucratic structure where, for a variety of reasons, it is often difficult to elicit support from the membership. Many unorganized workers are ambivalent about the benefits to be gained from belonging to a union. In this context of labor-management relations, organizing around health and safety issues is often a frustrating experience. Thus, some activists "burn out," and some opt out by leaving the job. Others continue at a level where they can deal with the strains. A few are co-opted by management into supervisory positions. And some find opportunities to move up within the ranks of their unions.

Thirty-four of the people we interviewed described themselves as activists. In this chapter, they describe their efforts to seek changes in the conditions of work. When they recognize a problem in their workplace, they initially follow existing grievance procedures, contacting their supervisor or higher management, and using union channels where available. When these procedures fail to produce results, they are faced with making a decision—to accept the situation or to pursue the problem, often at considerable personal cost.

### "It's My Style"

We asked activists why they spend so much of their time and energy trying to change a situation while others have tended to adapt. Some attribute their activism to peculiarities in their personality: "I'm domineering," "I'm just outspoken." Others describe the influence of their families; they were raised to be "independent," or "to do something rather than complain," or "to question authority." They feel that activism is part of them, a fulfillment of their values, a way of life.

---

### Mike, photo lab processor, blueprint company

I could tell you what I call the people I work with. I call them jellyfish or ostriches. They just hide in the sand. They don't want to get involved. They'll gladly take the benefits. They'll sit there for the popcorn but they don't want to shake the pan. It's because people are like that that they have to work in the conditions they do. In our family, if you're going to gripe, do something about it. If you ain't going to do something, shut up. If I don't like where I have to work, I say so. Why shouldn't I? I won't say I'm a troublemaker. I don't go out and blab all over the place, but I'm not afraid to talk to somebody about safety conditions here. If you look at these chemicals and what they can do to you over a period of time, they're downright dangerous! A lot of them don't take effect until 10, 20, even 30 years later, so you don't know what's going to hit you until it does, and then it's too late. I figure, we're not just talking about getting sick. It's our lives.

### David, chemical operator, pharmaceutical plant

I have always been very outspoken. Why? Just my feeling of right and wrong. I hate to see people get walked on. Somebody has to take charge. With some people—if somebody does something to them, they just lay back and take it. If nobody is there, they'll just lay down all their life. Somebody has to take a firmer stand on hazards. It's people's lives that we are dealing with. But I'll probably get all hell. It's why I am on a lot of peoples' shit list.

### Dick, granulator, pharmaceutical plant

You could say that activism was kind of born in me. It started with my father. He's been an active union person for a long time. He was in the merchant marines for 23 years and in the utility workers after that. He was president of his local for nine years and now he's on the Executive Board of the national, so it kind of rubbed off. I just didn't feel like being a worker there and letting everything go by. I have to have a say in some things that go on.

### Gene, pipe fitter, chemical plant

I guess I'm so active because of my self-confidence and arrogance. What else can I say? I'm very domineering on the job. Once in a while I have to back off. It's just my character. I was raised to be very independent and self-sufficient. I left home when I was 14 and have been on my own ever since. I hated my stepfather with a passion while growing up, but as I got older, especially once I got away and in the service, I realized he taught me some valuable lessons that would last a lifetime. I might not have appreciated it in my teens, but I sure appreciate it now. He raised me to be totally self-reliant, and have the character and assurance that when you decide this is what you're going to do, you do it. Once you start something, go all the way through. I think a problem with a lot of people today is lack of self-pride and self-respect. If you don't have that, what have you got?

### Nora, graphic artist, blueprint company

You know what it's like for a woman in a lab to be standing alone and have maybe seven or eight company men across the room from you? You know how intimidating that can be? Faith helped me get through times like that, when it would have been easy to drop the whole union thing, to say forget it, to just turn my back. I drew a lot of strength from my religion. If the company knows what they're exposing people to, it's immoral, it's just wrong, there's no other word for it. I can relate that to my faith.

### Tom, print machine operator, university

I go along with the bumper sticker that says, "Question Authority." And that's essentially where I stand. I'm a radical thinker. I believe that no one should accept the final word on anything. I don't agree with Reagan that business is the backbone of the country. I don't believe that. I believe the average worker is the backbone of this country, and the more that could be done to educate and to protect that person, the better things will be. And it's got to be done now.

## "I Got Radicalized"

In explaining the source of their activism, a number of workers described personal experiences which radicalized them. In some cases, this was a political event; in others, activism developed from specific experiences at work. Health and safety advocacy became a means to turn their anger to constructive ends.

### James, computer assembler, manufacturing plant

The Navy did a good job of making me an activist. I was in from '68 to '70. I went in feeling all gung-ho. I wasn't quite sure whether I wanted to be gung-ho in Vietnam or not, but I didn't think I was going to refuse to go. All kinds of things started happening there—just the way the navy was treating people. I tend to be a democrat at heart, not a Democrat as in party, but with a small "d." If we were supposed to be fighting for democracy, why weren't we being treated the same way? Then, I turned against the war. I was in the service when Kent State happened, and it just turned me around. I came out of the service and went to school wanting to become a history teacher. I always liked history as a kid, used to read a lot about it. As I read more and more on my own, I started questioning things. It opened up all kinds of avenues. I started making these webs, seeing things in a different light.

I got out of school and worked in sales for a while, but got tired of ripping people off. The name of the game was sell, sell, sell, profit, profit, profit. Disillusioned with that, I said, "Well, I'll go to work in a factory. At least I won't be screwing anybody." I decided to go to one of the better ones—here. They put me on assembly and I left my radical college days behind, so I thought. But then I started feeling more and more like this is just another rip-off. Here I'm trying my best and they're harassing me for overtime, and exposing me to all these chemicals without telling me what they are. Back again to that small "d" democrat. I told them, "Listen, this isn't right. We should be able to challenge and to know." They don't like that kind of talk and that made me even more pissed off.

I didn't know much about unions or occupational health and safety. I went out on my own to find information and to write up leaflets. I started bringing more and more people into the group who felt the same way. Ironically, I feel like I have to stay in this plant in order to organize

the place. It's like I'm caught between a rock and a hard place, because
I do want to organize it, but I have to put up with all this other shit. It
makes you wonder, will it be just frustration all down the line or will
something good finally come out of it?

### Steve, railroad trackman

I feel extremely angry at some of the day-to-day conditions we're
all subjected to. So when I had an opportunity to do something, I grabbed
it. The first time I was fired by the railroad was because I became, some-
what rashly, a self-appointed shop steward and safety man. Together with
a few other guys, we decided we were going to do something about safety.
So we slowed up a job by removing certain unsafe, badly worn tools. The
reaction of management was semi-hysterical. They sent the supervisor
out to watch and the next thing I knew I was put out of service, charged,
given a fair trial, and shot, so to speak. Five weeks later, the union got
me back to work, with a little helpful advice that I should keep my mouth
shut. I never got over the fact that I had lost about $1000 for merely doing
what I thought I had every right to do.

Well, it's a lot better now. I became an official safety representative, and I really locked horns with management. It was rough for a
while. But I have the organization behind me, and some official sanction
and immunity. By now it's a way of life—protecting the safety of the
people I work with.

### Ben, repairman, chemical plant

Why am I active in safety? Partly because I kept getting hassled.
You can't wear your safety glasses because they're all steamed up, but
if you take them off a piece of the pipe may blow in your eyes and they'll
blame you for not having your safety glasses on. Then there's an argument:
"I couldn't work because the steam was steaming up my glasses." "Then
what did you work there for?" "The foreman told me I had to do the
job." "He never told you to take the safety glasses off." "I just took
them off because I couldn't see." And so on and so on, a big routine.
That's what got me into safety.

I got involved in the COSH group.[1] I think it's an excellent
organization and I wish it would do a lot more than what it does right
now. The people in it go out and get the education and pass it along not
only to labor but to management too, if they would only recognize it.
Through the COSH I started learning about some of this stuff that could
be dangerous. It bothers me now to see a guy standing in a bunch of dust

---

1. COSH groups, Committees on Occupational Safety and Health, are regional coalitions
of workers, trade unions, and health specialists that have organized throughout the country
to assist and educate workers.

and come out gagging. I ask myself, "Why is he working there to begin with? Why doesn't he put a mask on?" I used to work with those masks and they're not too comfortable, but now I know enough to wear a mask rather than come out looking like that. I used to be one of those guys who think, "Well if I get it, I get it. Too damn bad. If my number's up, my number's up." I guess it's typical for most men. I found out that's not the way to think. If there's a problem that can be eliminated, eliminate it for the next guy. If we don't, before too long there won't be a next time. I feel that my father and his forefathers kept it going up to my time: now it's my responsibility and yours and anybody else out there to keep it going for the kids. If we just let things go because of profit or enjoyment, then we're not really doing our job as we should as human beings.

---

### "No One Will Do It for You"

Activists convey a vision that is consistent with American values of self-reliance and democratic control. Rejecting the paternalism of government or employer, they emphasize the importance of independence: "People have to help themselves"; "No one will do it for you"; "We have to solve our own problems." Self-reliance sometimes takes the form of individualism; more often it is a collective goal aspiring to industrial democracy. The point for the health activist is that participatory control is necessary when workers cannot rely on anyone but themselves.

---

### Sandy, rigger, chemical plant

After work I go home and sit at the typewriter and write letters to senators and congressmen and send out fact sheets and health hazard alerts. I guess I'll never be able to relax and enjoy myself, like some people can, until the issue of worker safety and health is resolved. I see workers themselves learning about the problems faced by workers. I see unity, I see a coalition of workers for workers. I see the existence of a feasible and viable Occupational Safety and Health Administration, or whatever agency, to ensure worker safety and health. I would like to believe the theory that Big Brother, the employer, will take care of you. But workers have to realize that they have to take care of themselves. If you don't do something for yourself, then nobody's going to do it for you. And that holds true in safety and health. My wife is very upset about my working with all these chemicals. She would like me to quit. But if I were to do that and obtain another job, I would be just throwing in the towel, giving up that commitment to a safe and healthy workplace. I'd be reducing my own risk of occupational disease or accidents, but doing nothing for the workers who remain.

### Ed, maintenance mechanic, university

I never think about calling OSHA. We have been individualists on our own for so many years that it's kind of hard to think that somebody else is going to fight for you. This is why I'm getting active in health and safety. I've been trying to convince the boys in my own shop that we've got a union to fight for us now. Before, we never had anybody to fight for us. We always solved our own problems.

### James, computer assembler, manufacturing plant

The younger workers today are aware that if you want to live to a ripe old age, then you're going to have to be more concerned about what you're working with. A lot of the older workers still have the old attitudes. Most of the chemicals we work with were developed within the last 20 years. They really don't understand them. It's like any new-fangled gadget. You have people working in the place for 40 years, and they're going to retire pretty soon. To them it's just another chemical. Where you see the changes are with the 30-year-olds who are just starting out their family life. They've seen people get sick from working with chemicals and don't want to end up that way themselves. What good is having a family if you're going to die when you're 40 or 50? They say that 80 percent of the people don't like the jobs they are doing, and you find that here too. Then why the hell should you let that job make you sick?

### Eve, sorter, manufacturing plant

The more the economy goes down, the more they lay off people. They no longer have the people to maintain equipment like they used to. We have to insist that they do. If the people would only refuse to work in dirty areas, they would have to clean them up. But some of them just say, "Oh well, they'll probably have it cleaned tonight." Well, don't wait until tonight. Tell them to clean it or else, because they can't fire you if you refuse to work in something that you know is hazardous. People talk about health problems but do nothing about it.

As for me, I'll go right into the guy's office; I don't monkey around. I don't know if he likes me to come in or not, but I'm not afraid of much. I worked in a shoe factory with no union and feel that in a factory with a union you shouldn't be afraid. I've never been afraid of foremen or anything, even when other people seem to be. I've learned the hard way that if you want anything in life you have to fight for it yourself. People say to me, "I wish I could be like you." I say, "Lord forbid. If they had everybody like me, they'd probably close the doors."

*Arnie, chemical operator, food processing plant*

After being out of work for six months, and being a high school dropout, I realized there weren't many jobs around. You either took a job at minimum wage in a warehouse where your health deteriorated because you couldn't afford to feed yourself, or you took a job in this place at a wage you could live on but where you subject yourself to all kinds of hazards. I told myself, "You know, this is the reality, so you better fight to change this place and do something about it."

*Gene, pipe fitter, chemical plant*

I got a little record book, and every time I got fumes or needed medical treatment I wrote it down. Within six months I had some 30 accident reports. I kept records of what transpired, what built up to it, what the times and lapses were. The vice-president used to come around to the workers and ask them about their problems because the product we worked on was one of the big money-makers at that time. When he was talking to me, my supervisor cut into the conversation and told him that, of all the accident reports in the plant, half of them were mine. I went over to my locker and got my book out which showed how I tried to get the supervisor to correct the situation. Later the supervisor apologized. It amazed him that anybody kept records. It's reached the point where you have to keep your own records. The only kind of safety you have is yourself. If you don't use common sense and the knowledge that you've picked up over the years, you're going to get hurt because you can't rely on anybody else.

---

### "A Better Life for My Kids"

An important motive for many activists is the hope that their agitation will help their children lead better lives. Some want to create the conditions to maintain their own health so that they can provide the money to give their children the education necessary for a "professional-type job"; "I'll try and give them a better life than I had." Others hope that union organization around health and safety issues will improve the conditions in which their children will someday work.

---

*Ben, repairman, chemical plant*

I don't want my kids doing this kind of work. It's too hazardous. I really wish that nobody would have to do it. I don't think it's a job that you can take pride in. There's no incentive for you to go anyplace, and after a while you ask yourself, "What did I get out of my life? I worked in a factory, I worked for my kids to make them happy and to send them to school, but actually my life is nothing.'' I wouldn't want my kids to go through that.

### Jerry, materials handler, glass factory

People come in to do their work with the attitude, "don't bother me, I'm here to do a job." There's a selected few who carry the ball for everybody else. You're talking probably 50 out of 800. The ones who do the most complaining are the ones who won't come forward and say anything in a safety meeting. I'll speak up because I just don't like taking risks. I don't like doing things that will hurt myself or other people. I'm the only one working in the family, and I have the responsibility to provide for my wife and two kids. So I can't afford to get hurt. I expect my kids will have a better life than I will. If it's at all possible, I'll try to give them a better life than I had, provide them with a better education, and things like that. Hopefully I can give them the means to get a professional type job where they don't have to work in a factory like I do.

### James, computer assembler, manufacturing plant

I've always said that I'll never let my kid work in this place unless there was a union really looking out for him. I have some deep problems with the industry. Not that I'm going to go blow up buildings but I wonder if the risks are worth what we're making. We're using these chemicals to make computers, chips, and information systems. In the long run, is it really worth it? Aren't we risking the possibility of polluting everything in this country? The groundwater supply is already at a dangerous pollution level. Workers are getting more and more chemically contaminated. If they keep going that way, I don't want my kid in there at all. Why would I? It's almost like telling him to go walk on the freeway. I would hope he'd get into something else. Hey, I'm sounding just like my father. He's a salesman; it was a really dog eat dog job and he always wanted something better for me.

### Rich, orchard worker

I've been active in getting a union, hopefully to make some jobs less hazardous. It might make things a little bit better for my kid so he won't have to deal with that kind of stuff. He'll have millions more chemicals when he's old enough to work. Hopefully, he'll have a way of dealing with it and he'll have an employer who will supply the right equipment and training. Chemicals are here to stay; they're a threat, but they're also a boon. You got to deal with them. If pressure could be put on employers to not go with the cheapest but the safest, that will be to my kid's benefit.

### Ted, welder, chemical plant

I don't hate the place because they've paid for my home and my family and I've seen the benefits. It's my lot. But I do hate what the work does to some of those people I've worked with, and I don't want it to do it to me. And I sure don't want it to happen to my son, or to his children. I feel responsible for that. Things have gotten better, and I feel partially responsible for some improvement. Some of those people, if I can speak frankly, are rotten bastards. They've knowingly killed my brothers and sisters in the union, and not cared. I have a responsibility to see that it's a safe place to work. I don't want to leave it the same way I found it. I hope it will help my son. I want him to be his own man, an individual who doesn't have to take bullshit from some idiot who doesn't know what he's talking about. I don't want to see him get into that kind of rat race. I want him to be able to decide, "This is what I want to do, this is good for me, this is good for my country, this benefits society."

---

### "Activism Has Its Risks"

Activists often work alone with little support from their co-workers. They seek information about workplace hazards. They appeal to OSHA. They directly confront management and try to organize people to change hazardous conditions. In these efforts activists risk harassment by management which can threaten their job security. They risk conflict with friends and family, who resent the demands on their time. They risk isolation by their union and co-workers, who fear that activism may jeopardize jobs.

---

### Vivian, laboratory technician, research institute

There are only a handful of us who are active in safety and health and making people aware of the problems. We're subject to harassment—they think we are instigators, troublemakers, nothing is very explicit, but we can sort of feel it. It's an undercurrent within the building. People sort of say, "Well, don't associate with them because they cause trouble." Most of the activists have left.

### Arnie, chemical operator, food processing plant

When I was elected to the union bargaining committee, the company thought I was a Communist. I mean literally. When their plant in Ethiopia was nationalized by the government during the revolution, the vice-president called me up to ask if I had any advice. He thought that I had some sort of hot line to China or Moscow or something. I couldn't believe it. I didn't know Karl Marx from Groucho Marx.

### Steve, railroad trackman

There's a chemical plant on Long Island and they were on strike a few months ago. My local supported them on the picket line. I was on the picket line one day and a scab crew, which was working the plant at the same time, went out one windy day and sprayed some herbicide along the siding so that it blew directly over the picketers. Several of us got violently ill and had to leave. Who knows what this herbicide was, but it laid waste to the vegetation as effectively as the phenoxy herbicides. It made us sick. They told me that the stuff had been stored in barrels for years on the property. Putting it all together, was it 2, 4, 5-T being used illegally as a direct weapon against picketers? I saw the results. They sprayed for about a hundred feet right on the railroad mainline. Was it 2, 4, 5-T? It might well have been. Who's going to know the difference? Who's going to do anything about it?

# PART 4

# Recourse

# 8    If There's a Hazard . . .

What do people do when they identify a hazard in the workplace? Where do they go for help? How workers deal with hazardous conditions often depends on the size and structure of the enterprise, management attitudes, and the existence of a union.

Under the OSHAct and the National Labor Relations Act, employers have the responsibility to respond to workplace hazards, and most firms have an established structure for reporting and correcting problems. Unions, as part of their "duty of fair representation," likewise must take actions to protect their membership. According to the Quality of Employment surveys, 84.8 percent of those people who report hazardous working conditions go directly to their immediate supervisor or to management. About 7 percent go to the government, and only 5.6 percent of union members report problems directly to their union representatives. However, 51.1 percent of workers facing hazards fail to report their problems at all.

In this chapter, workers talk about their experiences in taking complaints to management, their attitudes about their unions, and their views on the means to resolve disputes.

### The Company's Response

Whether or not workers report problems often depends on the receptivity of management to complaints. Some firms encourage workers to identify hazards; others appear to discourage communication. Once problems are identified, responses vary. In some companies, reluctance to fix problems seems to increase

**113**

with anticipated costs. Other firms resolve problems promptly.

Many of the workers we interviewed feel frustrated in their efforts to call management attention to health and safety problems. The difficulties often stem from their relations with supervisors. They feel that supervisors belittle or ignore their problems; that the imperatives of production leave managers few incentives to deal with health and safety: "They bury our complaints." Workers in larger plants sometimes have recourse to a health and safety committee. Such committees, composed of management personnel from different departments or some combination of labor and management, review problems brought to their attention and, in some cases investigate conditions. Although they are usually limited to recommending improvements and rarely have independent authority to make changes, many workers view them as a useful recourse.

---

### Walter, pipe fitter, glass factory

The company's running the place, so I feel they are responsible for hazards. It's just like you coming into my house. If I've got something in here that's going to hurt you, it's my responsibility to tell you or I'm liable for it. That's my feeling.

### Lyle, painter, chemical plant

When we complain, they have that "hide-all" attitude: "Engineering is working on it and when they come up with a solution we will take care of the problem as soon as possible." But it seems like somewheres along the line problems are buried. Unless you keep complaining, you won't get any satisfaction. There are times when they just plain say, "Look, this is the best we can do. The company's here to make money, and when we get around to it that's when it's going to be done." We once had a safety coordinator who I felt was honestly trying to bring about good safety changes in the plant. He made some changes, but after a while I guess he got to be a pain. Word came from above to knock it off.

### Greg, air-conditioning repairman, university

The management safety department is a joke. They probably know what to do when somebody gets hurt, but when it comes to these chemicals they know less than we do. Once when the plumbers were treating some lines, they mixed some sulfuric acid 50-50 instead of 20 to 1. After they were done, they called the safety department to get rid of it. The safety guy looked at it for a while and said, "Well, you can either pour it down the drain or send it to Maryland. What do you want to do?"

This is 50 percent sulfuric acid! And it was just sitting there in a garbage can in a dormitory with no markings.

### Joe, laboratory assistant, chemical plant

If you report a hazard, it's not taken care of until something happens. For example, we reported a cyanide smell. I don't know how many times we reported it and nothing was ever done about it. One day we came in at 8 o'clock in the morning and shut the equipment down. Rather than repair it, they let us sit for eight hours and waited for the next shift. At 4 o'clock when the next shift came in, they started it up. This went on for two or three days until a guy got knocked out from cyanide poisoning and was taken to the hospital. Then they finally found out what it was. One of the reactors on our scrubber system had a good eight to 10 inches of the elbow blown out so there was no suction on the cyanide at all. But they waited until someone got injured before they even thought of looking to see what was wrong.

### Lee, stage carpenter, university

If someone falls, there's bones on the floor and everybody can see that cause and effect are clearly connected. It's much easier to get them to spend money to correct the problem in a case like that than when you say, "I'm not sure we should have acetone in here because I get headaches after about an hour." Once we had a problem with the ceiling beams in the main theater. We were worried that one of those beams would fall into the audience. That they didn't ignore. It's one thing to kill your actors and actresses or your technical crew, but quite another to attack the public after charging admission.

### Fred, chemical operator, chemical plant

We had a strike here three years ago. I blame the company for the strike. We wanted more say about safety in the plant. They said, "It's our damn plant and you ain't telling us how the hell to run it." We weren't going to tell them how to run their damn plant even though I think there's a lot we could say about it. They fight us tooth and nail. We used to be able to work things out through the grievance procedure. Now—no, no, no, no, no, on everything. Their whole attitude has changed to the point where I'm looking forward to getting out of here. They spend money, but only to improve production. For safety alone you have to fight with them. They'll sacrifice safety just to save a few lousy dollars.

### Ted, welder, chemical plant

We had a caustic spill on the road. We got the cops and newspapers and photographers to look at the stuff. Nobody knew what it was.

Then the company sends this research chemist out. He gets down on his hands and knees and sticks his tongue in 50 percent caustic solution. He didn't show us his tongue.

### Irv, plastics fabricator, aircraft factory

We've worked out a routine that seems to work most of the time. The person with a problem will go see his immediate supervisor. If the problem isn't rectified, he then goes to see both the safety rep and the union steward, who will then go back to the supervisor. The supervisor has two shots at correcting it. Then it goes to management. From management it will be transferred to the company safety man, who has the responsibility to see it's taken care of. As a union officer I can't think of any problem that was of such major proportions that it had to go all the way to arbitration. We have safety grievances, don't get me wrong, but we deal with them across the negotiations table, not through the arbitration system. We have our monthly safety inspections, run by both union and management, and nine times out of 10 the company responds to our requests. Our big safety problems have always been taken care of before someone has gotten injured.

### Peter, railroad signal inspector

Someone read something that said that asbestos may be hazardous to your health. Supervision was unbelievable in how they reacted to this. It was, like, we let the devil out of the closet. I'm sure that they knew about it, and you would think, "Okay, now that everybody knows, we'll do something about it." They brought in a team of specialists who put monitors and masks on the people for a day. This was supposed to tell how much asbestos they took in. After all this was done, they said there was absolutely no danger at all. Zero. They came up with numbers. You can't argue with that because you don't know. You have to believe the people who did the testing, but they were hired by the railroad. The railroad says the material has only a small amount of asbestos. But it's grayish white and it has flakes. As far as I can see, it's pure asbestos. How they got away with it, I don't know. The safety of the men is the least of their concerns. There's a standard joke: "If you're working with a piece of equipment, be careful of that. You we can replace. That we can't."

---

## The Union's Role

Unions provide another important recourse in dealing with problems of occupational health. However, with some exceptions,[1] unions have been preoccupied with employment and wages so that occupational health is a recent concern. It was not until 1978 that the AFL-CIO established a department of occupational safety and health. For the rank and file worker, membership in a union may mean simply paying dues; typically only 5 to 10 percent of union members attend monthly meetings. Despite this minimal involvement, many workers see their unions as important in assuring their well-being, mainly through obtaining health and safety clauses in their contracts and by informing workers about chemical hazards. They are most concerned about the efforts of their union local, seldom referring to the activities of the international.[2]

Workers are often frustrated by the limited union influence over hazardous conditions. Preoccupied with bread and butter issues, some local officers regard health hazards as secondary. Union locals also lack the financial and technical resources to deal with complex technical health questions, having far less expertise than a company with full-time health and safety personnel. The international unions sometimes help. They have recently added health and safety experts to their staffs, lobbied the legislature and government agencies, and provided expertise to locals. Using grants from OSHA's New Directions program, designed to develop technical competence among workers through educational efforts, the internationals have issued publications for members, organized training programs, and created data bases for identifying potential problems. Even with this help (which is now affected by reductions in New Directions grants), locals usually operate as independent entities, left to their own devices in dealing with the health and safety problems they encounter on the job. They do so mainly through influencing the language of collective bargaining contracts and bringing complaints to OSHA (see Chapter 9).

### Ben, repairman, chemical plant

Workers have to fight for their cause and they need the union to protect them and to make sure that the government laws and the company rules really work, and that one's not controlling the other. Sometimes the

1. For example, the UMWA has employed medical personnel since the early 1960s and has a model Health and Retirement Fund. The Oil, Chemical and Atomic Workers (OCAW) have had an active health program since the 1960s. Similarly, the United Auto Workers have employed full-time health and safety experts since the 1950s.

2. Some unions are independent, limited to one company or plant, and are often outgrowths of company-sponsored employee associations or "company unions." Other unions are national or international (usually U.S. and Canada) and consist of groups of locals within an overall structure. Each local, whose members work at one plant or in one region, is itself organized in an independent structure. The terms "local" and "international" are commonly used to describe the two levels.

government laws can end up hurting the workers. Sometimes the company policies are wrong. There's got to be a teamwork situation between government, management, labor, and the individual workers. You've got to have strong government laws that say this is what you've got to do because it's right. You've got to have a working atmosphere that people can relate to in a good way. You can't intimidate people into doing things that will harm them by offering them money. And you've got to have a union to say, "Hey, you can't do that, that's not right."

### Steve, railroad trackman

Safety on the job depends on workers banding together to counteract something the foreman is doing. That's often hard. It doesn't just happen spontaneously. I've heard union representatives say, "It's the fault of the men; they should know better than to let the foreman do such and so." That to me represents a surrender of responsibility, because workers depend on leadership and they look to the foreman to properly conduct the job. The foreman is the center, the linchpin. You can't expect a group of men just to ignore something that the foreman is telling them to do. You have to have some countervailing structure, organization, and leadership coming from their organization, which of course is not the company but the union.

### Kitty, industrial painter, university

Everything I've got I owe to the union, and that includes my self-confidence, what I've got of it, and my sense of self-respect. When you can get up every morning and look yourself in the eye in the mirror and say, "Well, I belong to local — , I'm a painter," it makes you feel good. Maybe I'm like this because it was so hard for me to get in and get accepted in a job like this. It's real important. If you feel like you have some control over your job, you feel comfortable in your job; if you're proud of it, it affects everything else that you do. Let's face it, it's 40 hours a week you're there, not counting lunch. It's a lot of time, and if you're gonna spend 40 hours a week doing anything, you'd better like it and have some pride in it.

### Ken, electrician, chemical plant

I've been a shop steward for a couple of months, and I see you don't change things too readily. There's a step above my supervisor and another step above him that seem to run everything. It's very hard to deal with. I get the feeling that people with an education try to make me feel beneath them. I try very hard to keep things on an equal level, and that means not cursing or acting like a moron. I try to act like an educated person. It's frustrating, partly because these managers are very antiunion.

These are the kind of people who think the unions are the downfall of the country. There's a lot of that feeling now. When I first came I almost had that feeling too, that unions cause more problems than they're worth. When you get older, you embrace the union. When you're younger, you feel you don't need it. But I realize now that without the union it would be unbearable here. We have to have a little bit of a say.

### David, chemical operator, pharmaceutical plant

People in the plant don't have a good feeling about the union. The local itself has not gone out to meet the people. To me, being a union officer is not a job where you can just sit in an office all week and do paperwork. People are the name of the game, and there is no contact with them at all. All the people see is their union dues coming out of their check every week and that's like the end of it until they get in trouble with a grievance or suspension. Then they'll see the union. The officers, especially the safety committee, should be out there all the time, not just in a problem situation.

### Dorothy, deckhand

The unions have health and safety activities, but not at all directed toward chemical dangers. It involves things like how long you can stand watch and under what conditions, or who has to pay if you miss a boat that leaves port early. They set up guidelines for a lot of things, which is a good thing because otherwise you'd be getting a worse shaft than you get anyway. But as far as I know, I've never seen any effort to control chemical exposures.

### Bill, fire fighter

I think that unions have to reevaluate themselves. They are certainly going to need to have an awful lot of input in the future or we could lose all the rights we have gained over the last 40 years. The unions have got really sucked in and lost sight of what they should be doing for their membership. They've responded to everything that management wants them to. I think we have to go back before the fat times and look at why unions came about. The leadership for too long has satisfied themselves with monetary gains for their people thinking that that's satisfying all of their needs. Unfortunately there are many other things affecting a union member that the unions could be working on.

### Eve, sorter, manufacturing plant

The union is active in the plant, but we haven't been able to get the officers really interested in health. Some of them claim they can't smell anything, and if men can't smell fumes they don't think they're really there. So we're in the age-old problem of convincing the men to check out these things, because if you just go to the union committee and say there are fumes, they'll just say, "We can't smell them." Well, I don't care if they smell them or not, they're there.

### Jenny, laboratory technician, pharmaceutical plant

The union's Health and Safety Committee was down to our place last week for a safety inspection. This is the first time they've been around since I've been there. They're supposed to come but they don't because we're the waste plant and we smell. People in different areas don't like to come down here because of the smell. Can't blame them, really.

### James, computer assembler, manufacturing plant

I want a union that will stick up for the rank and file. There's not too many of those anymore. We didn't get involved with some of the established unions because they're so undemocratic. I mean, when you've got a union boss making as much as the chairman of your company, you tend to wonder, "Whose side are they on?"

### Arnie, chemical operator, food processing plant

When the union came in we set down this three-year plan to get the company on safety and health. It was pretty hard to organize around the issue, particularly the health hazards. The stuff that we know now was not widely disseminated back then. We had to start with the things that were most immediate.

When we sat down to look at the injury and illness reports of the company, we found seven lost-time accidents all related to the steps. That's one tenth of the work force in a year and a half. So we decided that fixing the steps was the first issue to take on, because it affected the biggest number of people. It was obvious, I mean it wasn't like a hidden hazard. You'd slip on the fuckin' steps all the time. We found that every set of steps in the plant except for the two that came down from the offices failed to meet OSHA standards.

We confronted management and said we wanted the rise and pitch of the stairs changed, and we wanted the surface changed. We wanted guard- or handrails put up and a wash system so that we could spray off the slippery oil and grease. We warned them that if they didn't correct the stairs we'd bring in OSHA.

They weren't very sophisticated on how to handle a grievance. They told us, "Those steps only have to meet OSHA standards if you carry things up the stairs. So we're issuing a memo effective today, that no one will carry anything up the stairs." We called an emergency union meeting and got everyone to work to rule: "They say don't carry anything up these steps, we won't carry nothing up. Nothing!'' When we had to load a still, we didn't carry a pipe wrench up those steps. If we had to use a pipe wrench, we'd go and find a lift truck. Then we'd go and find a flat, put our pipe wrench in the middle, bring it over to the still, and haul it up. That all took 45 minutes. After a week the company caved in. We got what we wanted.

It was much harder for us to get them on the fumes because you had to prove there were so many parts per million in the air. But our credibility had been built up so it was easier for us to tackle some of these more difficult problems. Once we got the momentum up on the stairs, people started to realize, ''This is pretty fuckin' good.'' First we got the wash-gun system in and were able to use it to eliminate a lot of dust. Then we started going after the solvent fumes, the nitrogen oxide and dioxide fumes. We forced them to redo their entire extraction machine, which is where most of the solvent fumes came from. We estimated how much solvent they were losing, in terms of dollars and cents. When they sealed up the machine to reduce the fumes, that made the operation more efficient because it cut their solvent loss. So they were happy in the long run although they resisted it terribly at first.

---

### Grievances and Work Stoppages

The vast majority of health and safety complaints are resolved without lost work time. Very few complaints turn into significant disputes, and few of these lead to strikes.[3] To enforce contract provisions concerning health and safety, union locals use a grievance procedure.[4] This involves several steps beginning with appeals to lower management and moving through higher levels and sometimes to arbitration. Where the process fails to resolve a problem, locals sometimes turn to job actions or wildcat strikes. However, these actions are rare as a means of resolving health and safety disputes, where it is difficult to gain consensus about the need for action and where many members fear for their jobs. Because of the potential loss of pay, the threat of losing a job, and the likelihood of disciplinary action,[5] workers regard with-

3. According to the Bureau of Labor Statistics in the U.S. Department of Labor, in 1980 there were 3885 work stoppages involving 1,366,300 workers. Of these, 20 (.7 percent) involving 10,100 workers (.7 percent) were over safety measures, dangerous equipment, and other health and safety issues.

4. In some nonunion firms, company policy establishes the grievance procedure available to employees. In such cases, where there is no contract, the ultimate authority is usually higher management, rarely a third party. Disputes therefore focus on management policy.

5. Most contracts contain provisions prohibiting strikes while the contract is in effect, so if workers do walk out they do so illegally and are subject to disciplinary measures.

holding labor as the "ultimate" recourse, to be used only as a
last resort.

### Steve, railroad trackman

Last September they started spraying 2, 4-D on the job I was
working on. The smell was amazingly pungent. On the best information
available on that chemical, I closed down a job involving 40 workers as
soon as they started spraying. When it became clear that we were going
to take a firm stand, they stopped spraying, making all kinds of phony
apologies about not knowing what was going on. The moral of the story
is that a job action is the only way to protect ourselves. That isn't easy.
I personally had to conduct an on-the-spot education and agitation program on the subject, because I couldn't just walk down the track and tell
everybody to pack up and go without them thinking that I was flipping
my lid. I had to explain to them what the chemical was and what it had
done to our members and to the Vietnam veterans. I put the issue in these
terms: "Your welfare now and you and your family's welfare in the future
are at stake. They have absolutely no need to expose you to this; there's
absolutely no justification for doing it, and you have every right to stop
them." There was a high degree of unanimity about the action that we
took, and we won a modest victory. We got them to stop spraying near
us; unfortunately we didn't get them to stop spraying it entirely so that
innocent bystanders who aren't informed or organized will no doubt continue being needlessly exposed.

We got away with it. There were no repercussions and no one
lost pay. However there's a problem in stopping work. You stop even one
worker from doing one thing and you've done something very serious. If
you don't do it right, you can expose someone to charges of insubordi-
nation. Often workers themselves don't want to go along. Believe it or
not, in spite of what others might have you believe, workers come to
work to work. Really. And they may resent having jobs stopped by the
union unless it's very clear that their vital interests are at stake and there's
no other way. That's a special problem when the risks are not immediate
but long term.

### Gene, pipe fitter, chemical plant

If you file a grievance, you have to know what you're doing and
how to word it. You can't just tell them, "I'm not going to do that job."
Instead you tell them, "I want to do this job, but give me the proper
safety equipment." Then the union will back you 100 percent. If they
fired you for refusing to do a job when all you demanded was safety
equipment, they would lose with the Labor Board. So they won't push

the issue. I've done that. When I was chief operator I shut the whole process down until the supervisor gave me the proper safety equipment. We went around and around for about three hours. "What is the proper safety equipment?" "A completely enclosed suit." "We don't have one." "That's your problem, not mine." After a while management realized they had no choice.

### Lyle, painter, chemical plant

Before we went into a safety meeting with the old safety coordinator, we'd hash out an agreement. Sometimes he would be told that he couldn't do what we'd agreed on. During the meetings he'd say to us, "Well, I'll see you after the meeting," which meant his hands were tied, and he's telling us that we'll have to grieve it. Personally I'd like to be able to see things get done upon request, not through grievances. It's time-consuming. These are things that should be taken care of more immediately.

### Ken, electrician, chemical plant

Sixteen years ago, the union, believe it or not, was very, very strong. However, it wasn't interested in health and safety. They were strong in the sense they could shut the plant down. They have lost a lot of that power because now everything is handled by lawyers. That old threat of "We'll shut the plant down" doesn't go anymore, because somebody is going have to pay, either a jail sentence or a fine.

### Tom, print machine operator, university

I initiated a health and safety grievance on the questionable storage of chemicals. I requested verification of the condition of the chemical room and that's still pending. Meanwhile we're not required to go in that room. So far I'm satisfied with what has happened. I believe that we cut through the red tape because we placed the problem in the strong language of a grievance. That was effective.

### Sandy, rigger, chemical plant

We had a wildcat strike over a health and safety issue in 1978 because our brothers sustained repeated injuries. There was a problem going through the grievance procedure to rectify these situations, so the workers felt their only alternative was to wildcat. That's the only way that management will listen. Some of the conditions were improved. Others were forgotten after we returned to work. The company took disciplinary action. Employee Relations notified us that the company would give us time off because we needed to learn that there are procedures to go through in resolving problems other than a strike.

Sometimes you have to go to arbitration over health and safety. In some circumstances you have no choice. But I'm not for arbitration on health and safety issues, because when you go to arbitration it's supposed to be a 50-50 decision. And I don't feel that workers' safety and health is a 50-50 issue. It's either make it or break it. You're talking about workers' lives.

# 9    If I Call OSHA . . .

Government has regulated workplace health and safety hazards since the end of the nineteenth century. Initially, individual states were responsible for workplace conditions. By the turn of the century, most of the heavily industrialized states had regulations concerning the major hazards of the time, but their interest focused on problems of safety more than hazards to health, and the regulations were generally ineffective and poorly enforced.

Public interest in occupational health developed in the 1960s, when the rise in environmental consciousness increased union activity, and several workplace and mining disasters placed the issue on the policy agenda. Disappointed in the actions of individual states to protect workers' health, and recognizing their own limited ability to improve conditions, health advocates sought federal government regulation. Their efforts culminated in the Occupational Safety and Health Act of 1970, the creation of the Occupational Safety and Health Administration (OSHA) within the Department of Labor, and a supporting research agency, the National Institute for Occupational Safety and Health (NIOSH) within the Department of Health, Education and Welfare (now Health and Human Services).

OSHA's primary duties are standard-setting, and the inspection and monitoring of the workplace. Initially, the new agency adopted many of the voluntary rules established by industry, turning them into legally enforceable regulations. Both industry and labor attacked its early actions. Industry criticized the agency for being heavy-handed and nitpicking, while labor criticized its concentration on safety and its neglect of more serious workplace hazards.

President Carter's appointment of Dr. Eula Bingham, a toxicologist, as OSHA assistant secretary (1977–1980), brought substantial administrative changes during the late 1970s, and with them greatly increased public awareness of the problems of occupational health. Bingham emphasized the prevention of occupational illness. Under her direction, OSHA eliminated trivial safety rules and promulgated new health regulations, which won support from labor but active opposition from industrial interests. These interests emerged again in the policy changes brought about by the Reagan administration.

Reagan appointed Thorne Auchter, a Florida construction contractor, as the new OSHA assistant secretary. Auchter believed that, without government interference, business will take care of its workers at lower economic cost. In the climate of deregulation, he revoked or reinterpreted many of the regulations promulgated during the 1970s. Policy changes include: reducing the 1982 budget for enforcement by 15 percent over 1981, reducing the number of OSHA inspectors, responding to most worker complaints initially with a letter to the employer rather than with an inspection, exempting from routine inspections three-quarters of all manufacturing workplaces covering 13 million workers, and asking employers to evaluate Agency inspectors. The effect of these changes is suggested by comparing OSHA's compliance activity during two periods, February to November 1982 (Auchter), and January to October 1980 (Bingham). Under Auchter's administration total inspections were down by 21 percent, along with significant decreases in complaint inspections (by 32 percent) and follow-up inspections (by 72 percent). Citations for serious violations decreased by 33 percent, and for willful violations by 75 percent. Penalties for failure to abate hazards were down by 78 percent, and the backlog of worker complaints went up by 105 percent.

In this chapter we convey workers' perceptions of this changing governmental role in regulating health and safety and its effect on their access to government protection.

## Responsibility and Regulation

Although professionals and small business employees share a skepticism of regulation, most blue-collar workers strongly support continued government involvement in regulating workplace conditions. They feel that they have a right to government protection: "That's what we pay taxes for." They believe that the existence of OSHA is the only thing that prevents management from allowing conditions to deteriorate. However, they feel "sold out" by OSHA's increased cooperation with industry and efforts to reduce industrial expenditures.

---

### Tom, print machine operator, university

OSHA is what we have to work with, and OSHA is nothing. At the one health and safety training session I went to, I was alarmed when I realized how ineffective OSHA really is. The average American worker doesn't know this. I'd like to see an increase in OSHA activity. I'd like to see an increase in requests for OSHA inspections. I'd like to see more funds allocated to regulation, and I'd like to see more publicity about health and safety. There are documented cases of people who've lost their lives at their jobs, and that information is shoved into the background. People all over are being affected by these things and are too damned scared to do anything with the economy what it is. What we need is a system of government that says you can do whatever you need to do to protect yourself on your job. My fear is that without real strict government regulation our ability to protect ourselves will be suppressed in the private sector, and that individuals who pursue the problem will be persecuted and fired.

### Ann, silk-screen supervisor, museum

I think some sort of health inspector, just like the fire inspector, should come around and say, "You must do this." When a fire inspector saw that we had jars of paint in a makeshift wooden cabinet, we had to get big yellow cabinets that had fire stickers and heavy duty doors. They cost at least four or five hundred dollars apiece. To get something like that there has to be a law. It took me about a year just to get a staple gun that cost $14.99 at Sears. What we need is someone with some authority who gives the museum no choice and no loophole to say, "Well, this is good enough," or "They have masks." It has to be strict or else it won't be enforced. You really can't leave it up to undisciplined people like myself. I work with two other people and they have no more interest in taking care of themselves than I do.

### Dorothy, deckhand

I don't know the occupational health and safety literature, but it seems to me that this is a job with chemical exposures that no one knows anything about. No one knows anything about the health effects because seamen tend to be transient and difficult to follow up. There's an international law for protection of life at sea, but that law applies to how many lifeboats you have to have and how the hatches are to be wired tight. There's a lot of laws that apply to underwriting the insurance costs of the vessel. As far as I know, there's very little that applies to things like handling chemicals on the job. I don't know who should be responsible. Regulating ships at sea is nearly impossible. The captain is the boss, and there is no authority beyond the captain. That's it.

### Don, railroad conductor

I'd like to see some type of change in the law. If a child under 18 causes any damage to somebody else's house up to $1000, the parents are responsible for his actions, and they have to pay. If the government would only have some law like that with these chemical companies, then if something backfires or a disease occurs five or 10 years down the road, they would be legally liable. I think that a lot of manufacturers would think twice before dumping these new chemicals or drugs on the market, if they were legally bound and responsible for anything that could happen down the road.

### Earl, landscape supervisor, botanical garden

I'm for government regulation. We would not have the protection we have today for people using chemicals if it wasn't for government regulation. But the government funds for training pesticide applicators are gone. Programs like that are dead. Organizations like ours can do our own, but how about the gardener for the apartment complex across the street or the little crew that operates out of a truck and does lawns and gardens? They're going to do all the wrong things in ignorance. I'm not very expert about a lot of this regulation thing. I've read the complaints of industry and some seem reasonable. The burden is maybe too great in some respects. But I'm afraid that industry, in its own self-interest, uses examples of unreasonable regulation to remove an awful lot of regulation that protects us in our work. Regulations are very expensive to society, but so are the consequences, the illnesses, the lost work time. I'm not sure that deregulation is good for the economy in the long run. Regulation is just one of the costs of being a technological society. You must have these costs, and if you don't you have them in a different way, in a much more tragic way because they're human.

### Ken, electrician, chemical plant

These men who sit behind desks . . . maybe if I were in their position, I would also consider workers like little gnomes, people you don't want to know about. If it's affecting your livelihood or your money or your big house, it's different. The government, the people who set the standards for these chemicals, they say so many parts per million and it's perfectly safe. I can't picture any exposure to phosgene or cyanide as safe. You might be able to walk through the room without passing out but . . .

If the government gets involved, right away the company starts talking about shutting down the plant, and the workers get afraid of losing jobs. I'm at a point right now where I don't care. If I lost my job, it would cause me inconvenience but it wouldn't be a crisis. But I'm older. To a young guy raising a family it would be a disaster to be laid off. If there was something bad at the plant, he would say, "I'd rather work. Couldn't they just straighten it out?"

### Steve, railroad trackman

This government is obviously disengaging itself wholesale from worker safety. I think that those programs, organizations, institutions, and laws which have existed should remain in place. Their role is good and I support them. But the fact of the matter is that, at its best, OSHA was not something you could count on to protect the health and safety of workers. There were too many loopholes, too many soft, easy, co-operative OSHA inspectors. OSHA is no substitute for a union. With all the laws in the world, you still have to have that.

### Mary, housewife, wife of a railroad conductor

They have government agencies that protect you if you buy a defective washing machine or if your car doesn't steer straight, but nobody's interested in chemicals. You can write to Betty Crocker and get your money back for a cake mix if it doesn't work. You can sue anybody that sells you a house that sinks. But it's okay to put these chemicals on the market. It's okay to let people use them and not tell them what's going on. You just try to get one agency to listen to you. You can write letters and you can make phone calls for the rest of your life, but it's futile because industry is being protected by the federal government. Yes, we need legislation, but I don't know where to start. The little people like us aren't going to have any effect.

### Laura, filter cleaner, pharmaceutical plant

I think the government has a responsibility for health and safety. They give industry such a damn tax break that they should get involved. Of course, now things are working the other way, let business do what it wants and kick the government out. This could kill us.

### Mark, physicist, university

The whole notion of government regulation is that people aren't really supposed to understand what they're doing. One of my friends in school was an economics major and signed up for a class called "Regulation." He believed that it's standard for everything to be regulated. Isn't that going a little bit too far? I mean, being a Republican myself, I said, "Isn't it a principle of democracy that everyone is supposed to learn on their own?" The first case this guy did was on regulating the housing insulation industry. Big deal. Everyone who sees the Pink Panther on television knows what you're supposed to do. When it's wintertime you add six more inches and you're in the pink.

In my lab I have a famous cartoon: the government scientists are pouring saccharin into this rat who is completely bloated to 30 times its natural size, and one of them is holding its tail in his hand. The caption says, "No sign of cancer yet, up the dosage another 800 quarts." So I don't believe this whole carcinogen scare as much as some people.

### Ted, welder, chemical plant

They've known since 1930 about things like asbestos. The government should get off their butt, get into these places and make inspections more often. Hell, that's what we pay taxes for. Those people that know should have some responsibility toward us—John Q. Public—which is me, my children, and the people in the area. I don't want to put my political views into it, but as long as Reagan's going to spend his money on bombs, bullets, and bullshit, then to hell with the American people.

---

### Access to OSHA

Workers expressed their belief that OSHA no longer works in their interest. This belief is reinforced by the difficulties they experience in gaining access to OSHA inspections. The OSHAct guarantees employees the right to an inspection when they perceive a violation that may lead to physical harm or imminent danger. It gives employees the right to accompany inspectors and to talk privately with them. However, workers, especially those without the protection of a union, often fail to report problems and avoid contact with inspectors because they fear reprisal. While the OSHAct prohibits discrimination against employees who exercise their rights under the act, there is a two-year backlog of discrimination complaints. A worker who is illegally fired for health and safety activities must wait a long time for redress.

---

### John, maintenance worker, food processing plant

I wouldn't ever call OSHA. I mean, it's a good way to make your job impossible. If you really think something's wrong, better to figure out what it is yourself. I wish OSHA'd come in, but I wouldn't want to be the one who flagged the place. It's just that you wouldn't be able to work here after you did a thing like that. I once called the Labor Board on a job. I don't recommend it. Really. Like, say you owned a place. Business as usual. Then you hear some employee called in the government. That would piss you off, right? Even if it's just for some routine matter, once you bring in the law, the owners are gonna get pissed.

### Greg, air-conditioning repairman, university

I wouldn't dare call OSHA. Our jobs would be in jeopardy. We would be gone within a month. They would find some reason to get rid of us, because "you guys called in OSHA and you caught us with our fingers in the cookie jar." Might as well start looking for a job as soon as they found out you did that.

### James, computer assembler, manufacturing plant

The government should be involved in regulating health hazards and enforcing the rules. But when I've contacted OSHA, maybe twice, I didn't find them very effective at all. It seems like all kinds of things have to happen before you can get an inspection. We should be inspected regardless of what has happened. There should be enough investigators on that staff to make at least a yearly check on a company. So I'm disillusioned about OSHA. It bothers me that all this stuff is being cut by Reagan. Who are we going to fall back on, especially those workers who don't have a union to protect them? Our only alternative is supposedly our government, but it doesn't seem to be ours anymore, so we're going to have to find new ways of protecting ourselves.

### Pat, graphic artist, community agency

I always figured that when I got pregnant I would stop working here, because, not knowing the composition of the fumes, I wouldn't want to risk anything happening to the baby. The company that makes the equipment said things were fine, but I didn't believe them. So I called OSHA to check out the fumes. But OSHA wouldn't even deal with it, because they said I had to lodge a formal grievance against my employer for them to even come out to inspect. I didn't want to do that. I didn't have any grievance against my employer. All I wanted to do was find out about the fumes.

### Lee, stage carpenter, university

People don't think of calling OSHA, because there's a great deal of ambiguity about what regulations apply and what doesn't apply. I hear assertions that OSHA standards for exposure to hazardous material don't apply to educational institutions. Some people on the staff think that if OSHA ever came into this shop they would regulate us out of business.

They would make us build two-foot safety railings around the stage because there's a three-foot drop from the stage to the seating. We would
have to put our plays on behind this railing. This is the reputation that
OSHA has with a lot of people I work with, so they hesitate to make any
contact whatever with them. Personally I have my doubts that OSHA
could be coerced into even giving us an inspection unless there was a
serious accident. OSHA is so underfunded and understaffed that, if you
waited for them to get around to you, you'd be waiting a good long time.

### Steve, railroad trackman

We tried to use OSHA to protect us from herbicides, but OSHA
ruled on their own initiative that they had no jurisdiction over the railroad
on account of it being a government organization, which it isn't. NIOSH
did agree to study the herbicide problem. We sent them a Health Hazard
Evaluation request. They sent a team of investigators out and the railroad
made it look good briefly. That was the end of that. NIOSH wrote a couple
of letters to the effect that they didn't see anything extraordinary about
the health condition of our members' children. That was NIOSH's epidemiological study of the effect of herbicides. It was obvious that they
had no intentions of stepping on any feet in the government. See, it's the
government we're talking about now, because we're talking about Dioxin
and therefore about the Agent Orange problem. If NIOSH came up with
any disturbing statements about its effect on us, that would be like one
small weak branch of the federal government sticking a dagger in the heart
of the Pentagon, which is *not* a small weak branch of the government.
They, of course, are the culprit in the Agent Orange problem. Can a
criminal investigate itself?

### Dick, granulator, pharmaceutical plant

After three years we managed to get OSHA to come in to do a
three-day study, a biological sampling, while we were working with methotrexate. It took them three years because of a lot of red-tape baloney.
They were having an argument with the company over the validity of
urine and blood sampling. There could be traces of it in the urine, which
the company insisted would rarely mean that anything was wrong. They
kept putting it off so we kept agitating. Finally OSHA came in and took
blood and hair samples. They told the volunteers—there were six of us—
that three months after all the tests had been complete, they would go
over it with us. It's been a year and a half, and so far they haven't done
that.

### Inspectors at Work

A worker's access to OSHA is partly limited by agency inspection procedures. OSHA conducts some general inspections of workplaces under its jurisdiction. However, given the small number of OSHA inspectors and the many thousands of workplaces, it would take an estimated 75 years to cover all eligible sites. Thus OSHA began to target high-risk workplaces in the 1970s. Today OSHA will conduct only general schedule inspections at workplaces with very high lost workday rates. In responding to complaints, unless the problem presents a serious danger, OSHA first notifies the employer and requests a response. If the employer fails to correct the problem, the burden is on the employee to notify OSHA again. The procedure reduces and delays inspections. Once in the plant, inspectors do not have to inquire beyond the specific problem. Observing their relationship with employers and the tendency to anticipate and prepare for inspections, workers express a growing cynicism about the role of government in occupational health.

*Mike, photo lab processor, blueprint company*

We had an OSHA inspector come. I couldn't talk to him. The vice-president of the company followed him around everywhere. When the guy went outside, I put a note in his hand saying, "Read this when you can, soon." He stopped right there and read it right in front of everybody. If I had wanted to be so obvious I would have pulled up a chair and talked to him. I felt, "Oh, boy, my death certificate is signed. . . ." The guy wasn't too swift. I never got to talk to him. The boss was with him all the time. So what are you going to do? You have to sneak just for your own safety.

*Sheila, laboratory technician, research institute*

When OSHA came for an inspection, I was notified and asked to come down as the employee representative. I outlined what the problems were, suggesting some labs to go to and some people to interview. The compliance officer didn't interview anybody. He refused to ask for a list of chemicals that were being used. He didn't pursue the need for training. He didn't ask anybody if they thought they were using any unsafe work practices. He just went in and said, "This is what the problem is. We know it's an intermittent problem, but of course we have nothing to measure now because nobody's doing any of the work that caused the problem." Some people volunteered to recreate it for him, but he said,

"We can't do that, that's an artificial situation." So even if we know that an exposure is over the acceptable levels, the inspectors have to be there right when it happens. I wonder what happens when they've got a body lying there on the floor. . . . Does it mean that there was no violation if there's nothing to measure?

### Joe, laboratory assistant, chemical plant

I filed a complaint to OSHA about the bare electrical wire by the acid tanks, the cyanide leaks, the vapor leaks in the reactors, and the piss poor maintenance in general. When the inspectors came, I happened to be the one from the health and safety committee who was picked to go with them. The company didn't know that I was also the one who had turned in the complaint.

I didn't like the way they worked. The guys from OSHA were in the company office an hour and an half before I was notified that they were on the property. When I went over, they had already gotten the story. We went down to the plant, and when I showed them what was wrong, they said, "Gee, we didn't know about this and we didn't know about that. The company told us that all this had been fixed." I told them I was the one who signed the complaint, and I never saw them again. After that, most of the immediate problems like the leaks were repaired, but there were no fines and no warnings.

### Ben, repairman, chemical plant

OSHA inspectors come in and they look at something specific. If they'd just turn their heads 90 degrees and look down about 30 feet, they might see a jagged beam sticking out in the middle of the aisle, but they don't. I don't say that they overlook things on purpose. It's just funny how they look over here and not over there. I believe that if somebody put in a complaint in writing OSHA would do an adequate job. But they would come and do the specific complaint and never look next door. If they just come for a general inspection, they walk around and give citations on ridiculous things like the wrong type toilet seats, and then overlook major things.

The firm can make it look good when they have to. They have lawyers; they have clout. They know how to intimidate people. They know how to get the community on their side. If they know it's time for an inspector to come around, they treat their people like kings. Then, three months later, they let up again.

### Eve, sorter, manufacturing plant

The company always knows when OSHA's coming because when OSHA's coming we never have the same fumes as we have when OSHA's not coming. We had been having fumes right around the area I work in, but Thursday and Friday, when OSHA was there, the fumes disappeared. I believe the company shuts off vents to save energy and money, then when OSHA is coming they turn everything on. But how are you gonna prove it? I think OSHA should come in there with nobody knowing it.

### Lyle, painter, chemical plant

Whenever the company knows the OSHA inspector is coming around, there's a big push to go around and fix up everything that we have been trying to get fixed. As soon as the inspection is over, whatever violations he finds are fixed. Then you go and find other leaks, and it's the same old setup again. We have a good safety program going and some things are being done, but others are being buried. They always say that Engineering is looking into it.

# 10   If I'm Sick . . .

Workers who become sick on the job face two immediate problems: finding adequate medical care and getting compensation for time away from work and for the expense of being sick. In both cases, they face the inadequacies and inequities of a system poorly designed to deal with occupational disease.

The worker who seeks medical help for occupational illness faces a remarkable hiatus in medical awareness and professional expertise. Workers go to company-employed medical professionals because of company regulations, because it saves them money, or because they have no private physician. Some choose to go to their private physician as well, but often find that doctors in private practice are unfamiliar with problems of occupational health and see too few patients to draw meaningful diagnostic conclusions about work-induced complaints.

The number of occupational physicians is comparatively small. Comprising less than 1 percent of acting physicians, and usually employed directly by companies, they are relatively isolated from the mainstream of the profession. At most medical schools students receive only a few hours or less of instruction in occupational problems. Rarely are they trained to identify workplace sources of disease. Recent interest in the discipline, however, has led to occupational health and safety training programs in about 15 universities and medical centers, and in 12 NIOSH special training centers.[1]

1. These NIOSH Educational Resource Centers offer an alternative to industry training in the field. Recent cutbacks in funding may reduce their effectiveness.

Some 2300 occupational physicians and 20,000 industrial nurses are employed directly by industry, mainly in larger firms. Nurses perform much of the on-site delivery of services. Operating under the authority of physicians, they administer care for acute problems and are responsible for medical histories, minor examinations, health-monitoring, and record-keeping. Only rarely are they specifically trained to deal with occupational disease. The company doctor is responsible for preemployment and sometimes annual physical examinations, emergency care, and the evaluation and care of illnesses and complaints. Handling and interpreting data on the general level of health and safety in the plant, the physician or nurse in the medical department maintains records on individual workers and on the entire workplace. The doctor also handles workers' compensation cases, representing the company whenever there are disputes about whether an illness is associated with work.

Workers' compensation programs are designed to provide benefits and medical care for those with job-related injuries or disease. Prior to the establishment of workers' compensation, common law made it difficult for workers to sue their employers and win. Subsequent reforms of state laws facilitated workers' claims. In response, industry helped to create a system of workers' compensation that would substitute a predictable administrative program for the risky decisions of the courts. In exchange for their right to sue their employer, workers gained social insurance against workplace accidents without regard to fault. However, this system of insurance has many problems, especially in its application to occupational disease.

The compensation system was established primarily as a response to workplace injuries. The uncertain origins of many occupational diseases have biased compensation boards against those claims. Surveys suggest that only 2 to 3 percent of those reporting work-related disease, as against 38 to 43 percent of those reporting occupational injuries, receive workers' compensation. Of 1.8 million disability awards in 1975, only 1.7 percent were for illness. They are simply more easily contested than injury or accident claims.

Employers have a stake in contesting them. In most states they purchase their insurance from profit-motivated private carriers who want to minimize payouts. Large firms which insure themselves have an obvious conflict of interest in evaluating compensation claims. While a worker does not have to prove the company is at fault in a claim, the problem does have to originate in, or be aggravated by, the workplace. To keep insurance costs down, company managers and doctors often dismiss complaints as unrelated to work.

Finally, the adequacy of financial compensation is a continued concern. For example, in New York State workers' compensation pays only two-thirds of a weekly salary up to $215 in cases of total disability or death.[2] Partial disability pays a maximum of $105 a week. For many families this is less than a living wage, and does not cover time lost from work. These problems have led to numerous proposals for extending coverage and for new administrative procedures.[3] They have had little success. The costs of illnesses are essentially carried by workers themselves.

This is the context in which workers talked to us about their influence with company medical services and their frustration with the compensation system.

## Company Medical Services

The professional image of "company doctor," like that of sports doctor or military physician, is tarnished by the conflicts of interest inherent in the role. The concept of medical professionalism implies autonomy and total commitment to the patient. Yet company doctors are also employees (and sometimes stockholders) in firms committed to profitable production. In the eyes of workers, their divided loyalties leave company doctors suspect.

The profession defends its integrity and its primary allegiance to the patient. The American Occupational Medical Association developed guidelines for ethical conduct in 1976, specifying that "physicians should accord highest priority to the health and safety of the individual in the workplace, . . . avoid allowing their medical judgement to be influenced by any conflict of interest, . . . treat as confidential whatever is learned about individuals served. . . ."

The workers we interviewed perceive the company doctor (and often the nurse) in a very different light—as an agent of the firm, a pawn, a quack. Among themselves, they sometimes call the doctor "Mister" to denigrate his role. Identifying company doctors as part of management, they question their motives. They believe that company doctors are employed not to give care but to protect the firm against lawsuits and compensation claims, to reduce insurance premiums, to minimize sick leaves, or to deny the need for costly investments in health and safety equipment. In some cases they feel that company doctors trivialize the nature and extent of their illnesses just to get them back to work. They suspect that medical records and compulsory medical examinations are used against them. Thus, while the workers we talked to use the company doctors and take the required exams, they

2. This varies from state to state. For example, in California, compensation pays two-thirds of a weekly salary up to a maximum of $175.

3. The OSHAct set up a National Commission on State Workers' Compensation Laws. This group has delivered a harsh criticism of the workers' compensation system.

often do so with reluctance. They know that their medical records are the only evidence of symptoms that may later show up as occupationally related disease. But they also fear that such records may be evidence to screen them out of a job.

---

### Arnie, chemical operator, food processing plant

The old doctor that we had at the company was a very honest man. He told you if he felt you were being exposed to something. The first few times I went to him, he said, "look, you know, that place isn't a safe place to work. Be careful what you breathe." The company dumped him, and then they got this Jesus freak for a doctor, who is a fuckin' quack and a half. I mean, you walked into his office and you thought you were in some fuckin' temple. He had pictures of the Lord all over the place. He never told you nothing. The other doctor at least used to get back to you and tell you the results of your tests. You had some confidence that maybe he was telling you the truth. But this guy was flake-o.

### Joe, laboratory assistant, chemical plant

I went to the company doctor when I was getting these massive headaches every day. I didn't have my own physician because I never needed one. If I had a cold I just bought cough syrup in a drug store. So I went to him and he told me I had hardening of the arteries and gave me a prescription. I didn't know what they were, but I took them and still got the headaches. I was out of work for eight or nine days. Nothing seemed to be getting any better, so finally I said, "Screw this, I'm going to a neurologist." I explained the headaches and showed him the pills that the other doctor gave me. They were cortisone. He threw them in the garbage and said that I shouldn't be getting those at all, that I was too young for arteriosclerosis. I explained to him about being hit with the phosgene during an accident at the plant. He thought there was a very good possibility that was the cause, but there's no way to prove it.

### Greg, air-conditioning repairman, university

I went down to the clinic and they said that they couldn't find anything wrong with me. I asked them, "How come I'm passing out, and I can't stand up, and I can't drive, and my eyes are swirling in front of me?" They just said, "Well, have you had a cold lately? Did you go out last night and get drunk?" Then they lost the lab reports on a couple of our men who went down for blood tests. Just about the time we were really being forceful about this health and safety stuff, they happened to lose the tests. The guys were very upset. They said to hell with it, and went to their family doctors.

### Nick, chemical operator, chemical plant

I went to the company doctor and he's queerer than a three-dollar bill. As far as I'm concerned, he's doing me no good at all. Nor is their nurse. She prostitutes herself with the company by the way she handles things. All they're worried about is protecting the company. They make people come back to work before they should be coming back to work. They give exams to protect themselves. It's all a whitewash.

### Laura, filter cleaner, pharmaceutical plant

I don't go to the company clinic if I can help it. The only time I would go there is if I had an accident or if they called me for a physical. They're hatchet men, I mean, hatchet people—there's a woman there now. As a matter of fact, the company doctor is sick in the hospital and everybody is very happy because God works in mysterious ways. . . . They're not there for the benefit of the employees. They're there to keep you quiet and to put on a show for OSHA or the government or whoever. It's a farce. I trust my own doctor, I believe what he says to me, and I believe that he's working in the best interests of my health. The company doctor isn't doing that. In most cases he's a guy who has gone to industry because the pressures of private practice are too great. If you work for management, you just do what they say, get paid, and go home. There are some good company doctors; I don't want to be critical of all of them. But most of them forget their pledge as doctors and just take orders from management.

### Dick, granulator, pharmaceutical plant

I got bronchitis and went to my own doctor—he doesn't ask me what I'm working with or where I'm working. That's kind of foolish, because there's so much now that is related to occupational problems. People are sick, especially when they get older. I think the doctors have to direct themselves toward a person's occupation and what they're doing on the job.

As for the company doctor, from what I understand, he was an engineer that the company put through medical school. He's their pawn and he does what they say. He'll never back an employee. We're required to take a physical every two years. They say it's for our benefit. This year, in the contract, we're trying to change that. We don't want no physicals, and we don't want anything from them. If they want us to take physicals, we want to be able to go to our own physician and have them pay the cost. We want to have the medical people there for emergency treatment, but for nothing else.

### Bill, fire fighter

After the office building fire, we went to the hospital. There was a lot of apprehension, naturally. Even though we're exposed to chemicals every time we go to a fire, we don't always go to the hospital. The first thing the doctor said was to take a bath and don't use ivory soap—it will cause dry skin—and go have a couple of beers and wash it out of your system. And away we went. But first they did blood work. They did vitals and they drew blood—froze the samples. Unfortunately the blood samples got sent to a lab down in Tennessee and got all thawed out. The story is that they put the samples in the same freezer where the nurses kept their lunches. Every time the nurses needed room for their lunches, they took out our blood and left it on the counter, and when they took their lunches out, they put the blood back in. The samples were frozen and thawed, frozen and thawed, and ended up no good. So the only existing record of our exposure levels were destroyed. They would have shown the level of chemicals in our blood after the fire. Losing those samples not only works to their advantage immediately, but it works to their advantage in litigation at a later time. Well, it went from the sublime to the ridiculous after that. They didn't do exposure readings in the control room where we fought the fire, where our guys were crawling around in this crap. So we are never going to find out the concentration at that point.

We've heard so much bullshit. One doctor said, "Because it was hot in there things were probably destroyed; it was probably the safest place to be." Another one told us, "PCBs activate the enzymes in the liver. They may have been good for you." Then they did an experiment on some chicken embryos. They injected 12 embryos with the dust from the building, and three of them died. They said that losing three out of 12 was normal. What they didn't tell us was that the nine that lived were all born deformed. They didn't tell us that when they hatched they had five legs and two heads, and that they all died within a couple of days after hatching. In fact they didn't tell us any of this. It all came through the media when an investigative reporter from our local TV dug it out.

### Bob, fire fighter

I had some further blood tests taken about a month and a half ago. They sent them to our doctors. I haven't actually checked with my doctor because he's told me that they didn't do him any good—he didn't know how to read them. He didn't know what to look for. He said, "I don't know what PCBs look like in a blood sample."

### James, computer assembler, manufacturing plant

Out of curiosity I went to see the company doctor after the OSHA regulation on access to medical records came out. I told my manager I wanted to see my medical records, and he had to check with upper management first. Two weeks later the word came down, so I made an appointment. The doctor, he was really cold. I went in and he kind of half-hid the records. The main thing that I wanted to see was my exposure because I work with all these chemicals. But they didn't have that. He asked me why I wanted that, and I told him flat out, "Listen, if I come down with something in the future, I want it on record that I worked with these chemicals." He proceeded to tell me that there was nothing wrong with the chemicals at this plant, that in his 35 years of being a company doctor he had never seen an occupational-related disease. I knew better. The man was just flat-out lying to me. What choice do you have when you have to go to people like that? They're not going to help you. As a matter of fact, they'll steer you the wrong way. One woman went to this same doctor because of a rash on her arm, and the doctor told her she was holding her screwdriver wrong. I mean, that was bullshit! It was a chemical problem because she was working with methyl chloroform. So that authority figure, that company doctor, is telling people this, and they're figuring, "Well, he's the doctor, he should know." I think that's changing. I think we're seeing things as what they really are.

### Ted, welder, chemical plant

One fella had blips on his lungs and he has emphysema; he told me because I work with him everyday, but he wants to keep it a secret from the company because he's afraid that the company will use it against him.

### Eve, sorter, manufacturing plant

I wanted to do a job that was on the line and they told me I could. But then the foreman told me I couldn't. "It says in your file that you can't work with that black wax spray and acetone and triad [trichloroethylene]." I immediately went to the nurse's office: "What's this business that I'm restricted from working with this black wax and triad spray?" She said, "Well, don't you remember the time you came into the doctor and he took you off that triad?" "Yeah, he took me off of working with triad that was boiling, for a period of six weeks. He didn't say it was going to be indefinite, and this stuff isn't even the same. Let me tell you something, from now on when you people restrict me, you better tell me about it." They took the restriction off, but I got another job anyhow in the factory. From that time on I told everyone, "You better find out what they're writing in your medical record because I was restricted and didn't even know it. We should all be checking to see what they put in our medical reports."

*Jenny, laboratory technician, pharmaceutical plant*

I wouldn't take my dog to the company doctor. He's all company. No matter what's wrong with you, you go back to work. Under our new contract we don't have to go to him when we're sick. We can have our own doctor's opinion, then we get a third opinion. I try not to go to Medical, either. You get the slightest little scratch and you're supposed to go up there. I wouldn't because that's on your record from the day you're there to the day you leave the place; everything bad you do stays with you. Even the nurses are all company. There was a guy who was just retiring who had a spot on his lungs. They never told him until a couple of months before he retired. They don't have to tell you what's on your record. I believe that's wrong.

---

### Compensation

The frustration faced in seeking medical diagnosis and care is matched by the inadequacies of the compensation system in cases of occupational disease. Discouraged by the difficulty of winning compensation claims, many workers fail to apply for compensation at all. Sometimes they fail to recognize the disease as work-related; often they recognize the relationship but feel they cannot prove it. Many people do not understand the intricacies of the system, in particular the no-fault provisions. Some are afraid to file claims while still working at a company, fearing that they will jeopardize their job. Others are deterred by the long negotiations over the etiology of illness or the extent of disability. Designed for injuries and accidents, many state systems have a statute of limitations, precluding coverage of diseases discovered long after a worker has left the job.

---

*Ben, repairman, chemical plant*

I know a lot of people who have gotten sick, and a couple who died of cancer. I'm positive it's work-related. But how can you prove it? There are people right now who are finding out they have serious problems. I'm sure the company knows it's from work, but they'll say, "We wouldn't work with anything that causes problems." A guy was overcome by hydrogen sulfide and was rushed to the hospital. He comes back the next day. What do they do to make him happy? Give him a nice desk job. He sat at that desk for a week, putting papers in the files. Had his street clothes on. I told him, "You ought to get a form and turn it into the compensation board and get a case number on it. You told me that you had headaches. Tell them what happened just in case two years down the road you start getting sharp pains or something in your head. At least then you'll have a case number on it." But he sat over there, doing his paperwork. Compared to what he was doing before, it's a dream job. He's getting $10 an hour to file papers. That's typical, they do that to everybody. I even saw a union official who burned his feet in the shower because the water was too hot; the regulator on the thing broke and he got burned. I saw him in the office on light duty for 10 or 12 days because he couldn't walk. I doubt if he put in a form. He just went to the doctors, got bandaged up, and they kept him on light duty. He doesn't realize that it should have been reported as an industrial accident.

### Don, railroad conductor

When my daughter got cancer, she had to go for treatment every-day for nine weeks in a row. From here back to the hospital is approximately 100 miles. And with the gas, parking, and tolls, it comes to almost $75 a week. That's a lot out of our paycheck. We applied to I forget how many agencies. The American Cancer Society gives you a grant of $50 a year. If you're out of work on welfare you'd have no problem, but you're penalized for having a halfway decent job. There's nobody to help, so you wind up borrowing money. Thus you not only suffer the loss of the person that dies, but you suffer in the years ahead because you have to pay off the bills that were incurred at the time.

### Mary, housewife, wife of a railroad conductor

When our daughter was dying, all we thought about was what was going on. The last thing that we were thinking of was litigation. When you talk about burying your child, compensation is the last thing you really have time to think about. There was too much going on. It was a very emotional time, and you really don't think about anything but what you have to do, which was to get through each day.

### Kitty, industrial painter, university

A guy working with epoxy lost four pints of blood through his nose. He was in the hospital three weeks and nearly died. Then he sits across the table from me telling me that he was lucky, because other painters have gotten much sicker. He couldn't prove that it was the epoxy, so he couldn't get disability or compensation. It was pretty damned obvious to anybody that the epoxy had done it because he started hemorrhaging at work through his nose and had to drive himself to the hospital. They packed him from the back of his throat all the way up to stop the bleeding with tubes through his nose, and the whole bit. But he can't prove it was epoxy because there's no deposits in his body. I told him he should get a better lawyer.

*Stuart, mold maker, glass factory*

When J. got silicosis we bugged the company about screening the other guys to make sure they had no problems. The company finally got around to doing that, but I haven't heard any results. I specifically asked and they said they'd tell us no matter what they found out. Will they? From past practice, I'd say yes. But the time delay kills me. I'm not so sure it's their fault. The company doctor set it all up, so maybe he's sitting on the X-rays. Or maybe when he got them all done he sent the results to the insurance carrier. I'm sure they'd love not to have to pay this guy his disability. "Gee, nobody else's got silicosis. They've been there 30 years." I've heard them say, "If somebody gets hurt, it's their own fault." They just don't want to pay.

*Gene, pipe fitter, chemical plant*

A friend of mine died on the job. I worked with his widow: I went down to help her do different things around the house because she was stuck. A compensation lawyer approached her and asked her to sign a release giving him permission to get the medical records and dig into the case. He said she should have gotten some kind of settlement because he died on the job. It didn't cost her anything so she agreed to let him look into it. The union president at the time—he sold out to the company—he tried to bluff her by telling her that, if she didn't stop the investigation, the company would take back the life insurance and the other benefits she already received. She came to ask me about it, and I told her, "There's only one thing you should tell them to do but it's not nice. There's no way they can take the insurance back. That's a union-negotiated benefit. You've got it. If they hadn't already given it to you, they could play silly games to delay. But there's no way they can take it back."

*Ken, electrician, chemical plant*

In the old days, compensation was very easy to collect. It was almost as if they didn't care. They paid you and brought you back. Now, I don't know whether insurance costs have gone up or what. They own their own insurance company so they fight every case. They're using the union safety committee as a stooge to help them prove that the person trying to collect did something wrong. They can always point to something; they can always find a fault. They're not at all interested in discussing those things that could hurt you in the future but don't bother you now. They're only interested in why a man fell off a ladder. If they can prove he did something wrong, they're gonna give him trouble on his compensation.[4]

4. This firm will pay the difference between payment from workers' compensation and the worker's salary if the accident was the company's fault.

*Laura, filter cleaner, pharmaceutical plant*

They really are not so concerned about methotrexate causing damage to a female, but about its reproductive risks. They don't want somebody to have a baby that's a mutagen because of exposure to the product. If they know you were gonna have a miscarriage or a stillborn they'd let you work there, but they can't guarantee that the baby won't be born live and be able to sue them. Under worker's comp, the worker cannot sue the company and get what he deserved, but a third party, the child, can if there's any problems. So if that's the risk they say, "We're going to keep you out."

# PART 5

# Controlling Risks on the Job

# 11  Knowing the Risks

The quantity and complexity of information pertinent to coping with chemical hazards is staggering. An ideally well-informed person would know about potential hazards, methods of monitoring exposure, acceptable exposure limits, and proper handling and emergency techniques. Relevant information includes the individual's health status and past exposure levels as well as the collective health status of those who have worked in the same plant.

Most workers know little about the bewildering array of chemical substances they work with everyday. Many of those we talked to have access to material safety data sheets (see Appendix 4), but these are of limited usefulness because they rarely identify chronic hazards or provide monitoring or medical information. Some work with materials identified only by trade name; others have no idea of the generic identity of the substances they are handling or their possible health effects.

Workers often recount a variety of difficulties they encounter in getting information. First, technical knowledge about chemical hazards is generally limited. While industrial and government laboratories have tested the toxicity of many substances, they have focused attention on acute exposures and acute effects because of the urgency of these problems. Less common are investigations of chronic exposures because they are time-consuming and costly. Concerns about such effects are leading to more long-term studies. Still, the limited state of knowledge leaves much room for conflicting interpretation.

Second, access to existing information is constrained. Fearing that health and safety information would cause unwarranted anxiety or encourage refusal to work, corporate managers are often reluctant to disseminate it to employees. They also contend that disclosure of the generic identity of substances, or even detailed health information, could jeopardize trade secrets. Those firms which systematically provide information about hazards to employees tend to do so in abbreviated form.

Information practices are subject to few laws and regulations. OSHA regulations require that employers provide workers with copies of their personal medical records, exposure records, and information from the testing of substances used in the workplace, but they must do so only on request. In early 1981 OSHA proposed a standard requiring mandatory disclosure of information about toxics. Under this standard, all manufacturers would have had to label what they use in their plants. Later, under the Reagan administration, OSHA withdrew the proposal, reissuing it in the form of the "Hazard Communication Standard." This allows wide employer discretion about the information to be disclosed, again leaving the initiative to the worker. In the absence of federal policy, 11 states and municipalities have passed Right to Know laws requiring employers to provide information about toxic substances.[1] In these areas, we found that activists were well aware of the laws and using them. However, most rank and file workers are often not informed about their rights under these laws, or are afraid to exercise them for fear of losing their jobs.

The creation and dissemination of information about chemical hazards is one of the more sensitive and political issues in the struggles over occupational health. Controversy over this issue reflects the level of trust between management and workers. Management usually argues that generic identity is not necessary to provide employees with adequate training and information. Many workers, however, want to know exactly what they are working with in order to evaluate the risks themselves. In this chapter, workers express their concerns about information, their frustration in trying to obtain information in the context of technical complexity, and their efforts to find out what they need to know.

1. Right to Know legislation has been enacted in California, Connecticut, Maine, Michigan, New York, West Virginia, Wisconsin, and in Cincinnati, Ohio, Santa Monica, California, Danbury, Connecticut, and several other cities. Legislation was introduced in 1982 in Illinois, Indiana, and New Jersey. Some 20 additional states and localities are drafting and considering legislation. As an example of what the legislation covers, New York's law requires that employers post notices informing workers of their right to obtain information on toxic substances, to provide information when requested by employees or their representative on a variety of characteristics including identity of the substance, acute and chronic effects and proper use, and to provide initial and annual training to all employees who are regularly exposed to toxic substances.

## "We Need to Know"

Many workers want to obtain more information about the substances they work with in order to better protect themselves, to make informed choices, and to gain the background to seek changes in working conditions. Most people felt poorly informed, but a few, for example, the physicist and the dry cleaner, told us that they felt adequately informed and that they were willing to trust the manufacturers to give them appropriate warnings where necessary. This question of trust is critical in understanding the workers' need for increased information.

---

### Kitty, industrial painter, university

The paint cans list ingredients. They don't list what they could do to you. I've never seen a paint can that did not say "use with adequate ventilation." You don't think anything about it because they all say that, every one of them. It's a bit like, on the side of a cigarette pack, "dangerous to your health." I would like to know if something's going to do anything nasty to me, and I would like to know the proper way to handle it so that it doesn't. So many scary stories that you hear about people with mercury poisoning, lead poisoning, going sterile from chemicals in pesticide plants. You just wonder sometimes. The best thing to do is to know, to be aware of what it might do to you, so that you're careful; you make sure to have adequate ventilation and a mask so that you keep the damn stuff off of you and not breathe it in.

### Mark, physicist, university

We work with a lot of chemicals that, if handled improperly, are dangerous. There's one that's called Ecostrip 94, which we refer to as "Love Canal in a Bottle." There are lots of things there that are horribly poisonous; if you spilled them on your cat it would turn into a toad. I mean, we're all expected to be fairly intelligent. If it says on the bottle that if you spill it on your cat it will turn into a toad, you just don't go around spilling it on your cat. Like we use hot trichlor (trichloroethylene); if you stick your hand in the bath while it's hot, you'd come off with just a stump at the elbow. That's something you don't do. We're expected to educate ourselves.

### Earl, landscape supervisor, botanical garden

The training we have here is far above the standard for people doing the same sort of thing in industry, but we're a government employer so perhaps we have a greater sense of responsibility to protect the workers. We keep files of labels when we buy chemicals; we write to manufacturers for additional information. We sometimes telephone them with specialized questions. We've had a blight this year that affects just about anything and everything. We were advised to treat the soil with tetrachlor, which is a chlorinated hydrocarbon. It's not acutely toxic to human beings, but it's highly persistent in the soil. Who knows about its long-range effects? If that's in the soil and it's half life is 10 years in flower beds, people long after I'm gone will be planting and transplanting bare-handed in that dirt four or five times a year. So I decided against using it. We might lose all our annuals this year by that decision but still . . . I called the manufacturer. "Oh, it's not very toxic, it has a very high LD-50.[2] It's not particularly dangerous." Well, maybe it isn't. But I don't know if I'm happy with that. He isn't putting his hands in it for a week at a time to transplant tulips. I'm not sure if we know as much as we ought to about these things, and I've never seen any information about long-range problems. In a sense, the people using chemicals now are like guinea pigs. Their problems become a statistical record base. Eventually we might know something more, but it will be off of those who have been working with chemicals today.

### Ken, electrician, chemical plant

The company is terrified of somebody finding out their secrets. They went through some problems with trade secrecy when they were dealing with a cancer study, but they finally agreed to disclose information. The union worked out some sort of agreement not to divulge the information they received.

I would like to know more about this cancer business. I have yet to see a label or warning that mentions cancer. All it ever says is "hazardous" or "don't breathe vapors." Cancer's like a big bugaboo. I would also like to go into long-range invisible things like reproductive damage. My children are grown, but a young worker coming in would want to know that. It would give people some choices because there's a lot of bidding and moving around. It might not affect my job as an electrician but it could affect somebody else's right not to work with things.

### Jill, dialysis technician, health clinic

I'd like to know some facts on the dangers of formaldehyde; how widespread the problems are, what the symptoms are, and what can be done about it. Now I go complain, and they say, "What's the matter, you can't handle your job?" or, "You're too delicate," or, "Well, gee, nobody else has this problem." They laugh it off and blame me like it's my personal problem. If I got some information, first I'd study it myself and then I would get in touch with the people in the other seven units in this area

2. The dose that results in the death of 50 percent of the test subjects.

who work with formaldehyde and urge them to read it. Then I'd make an effort to convince our chief techs, saying that these are the facts, this is what we need, and let's do something.

*Eric, sculptor, self-employed*
    I prefer to know the worst that could possibly happen, exactly what the materials I work with can do. I just read a thing about additives in food and I was grateful to find it. It gives me some insight for making decisions about what I want to use and what I don't. I like details. Like with carbon monoxide, I like to know that the little molecule binds to your hemoglobin and prevents oxygen from going there. That makes it very real. It becomes a real danger. I need the specifics. They stay in my mind.

*Tony, dry cleaner*
    I don't know about people needing to be better informed. I think they get all the information they need. They need to know that this is a bad item to get on your skin, or this is a bad item to mix with another. These chemical companies are strong on letting you know all these things because they don't want to be sued by everybody in creation. Law suits spring up everywhere you can think of. Dow Chemical certainly would let you know everything that possibly could go wrong because, if things did go wrong, their reputation would be in bad shape. Cleaners all over the country would say, "We don't want any of that stuff." Not that the cleaners are going to ruin Dow, but that could be one source of trouble.

---

### The Problems of Getting Information
    People described the various problems they face in learning how to handle toxic substances. Most workers in large companies are given some kind of formal training; in smaller firms, they rely more on informal advice from co-workers. In both situations, they are upset by the inadequacy of training and the tendency to underestimate risks: "What they have is a public relations program."
    When companies do provide information, workers often found its usefulness limited by its technical complexity. They have to deal with language that is outside their ordinary experience: "I don't even know how to pronounce some of these things, let alone know what they are." They find it difficult to interpret scientific caveats and qualifications, especially when they receive conflicting information about toxicity and harm. We found that the complexity of technical information and the lack of resources

to use it effectively greatly contributed to people's anxiety and
their sense of alienation.

## Inadequate Training

### Don, railroad conductor

All the conductors have to take the Commuter Awareness Program. It's a "How to Be Nice to People" course. They went to a lot of
time and expense to get every trainman and conductor to a training center
for one day to show them how to handle unruly or intoxicated passengers.
If they would just make the same effort for a one-day class on chemical
problems, it could be very informative. Just like they teach you how to
take care of drunks, they should teach you about the chemical hazards.
But that's the problem right there. They're not going to pay you to go
listen to what they're doing wrong.

### Laura, filter cleaner, pharmaceutical plant

The monthly safety meetings cost them a lot of money because
everybody in the plant sits down for half an hour. They're still paid their
salaries, but then they only tell you, "We had one lost time injury, 40
minor injuries, and 2 near incidents." They never tell you who it happened
to or what happened so that you'd be more aware of doing something the
right way. It's the same in the respiratory training program. They tell you
how to put your respirator on and use it, how to change the cartridges,
and how to make sure it's clean. The people feel the program is a joke,
a farce, a lot of bullshit, but if it's gonna let them sit down for half an
hour they'll go.

### Ben, repairman, chemical plant

As far as having an adequate safety training program, what they
have is a public relations program so that they can say to you and to
anybody else that asks them that they have a safety program here. There's
a once-a-month safety meeting that consists of them showing you a movie.

### Rose, pill coater, pharmaceutical plant

I haven't seen anything on health come out from the company.
We have safety meetings once a month but it's never about chemicals or
anything involved. It's always driving tips or ways to keep your car
running in the winter—stupid things like that. They should be talking
about the chemicals we work with. But they don't. When I've asked about
some products, it took them seven days to get back to me.

*Arnie, chemical operator, food processing plant*

I'll tell you a story about the kind of training they give us. When I first started on the job I was a bulking operator. We did the bulking in huge vats. The tops of these vats were about 15 feet in the air, and we would have to dump these heavy 450-pound drums of vegetone into them. To do it, we used a small three-by-four skid. At the end of the skid was a slat that you could use as a lever to tip the drum. You put the drum on the skid, and you'd get on it too. They'd haul you up 15 feet in the air and you'd take this huge drum and tip it to pour the vegetone into the vat. One of the maintenance people finally said, "Hey, you could slip off that fuckin' thing, you know?" That was the training I got. Never once in the entire time I was there did the company give us any education on the health hazards of the work. We learned everything on our own. We relied heavily on a technician who worked in the lab and gave us information under the table about the solvents. If I had been trained more about the harmful effects of a substance, we would have had a lot more leverage in cleaning up that place a lot sooner, but we never got that information.

*Sheila, laboratory technician, research institute*

The lab director sent around memos all the time saying, "The way to deal with chemicals is to use common sense." Well, read me these 38-letter chemicals and tell me how to use my common sense. They should train technicians regularly, not just for the first week on the job. They should train the scientist who is really into his research and doing it for years. There are a lot of changes in safety procedures, and a lot of update on chemicals. The lab supervisor should take people around and show them the specific hazards they're dealing with in the labs—which means they should train the supervisors. God knows my supervisor didn't know what the hazards were. I mean, he's a guy who says, "I like the way this stuff smells."

### Technical Complexity

*Ted, welder, chemical plant*

The company will give us the information if it's required by law, but it's like handing a Stone Age man a rubber grip for his club. What the hell do I do with it? Where does it go?

*Ken, electrician, chemical plant*

They use letters; they call di-ethyl benzol DEB. Benzol is a very frightening name, so to call it DEB makes it sound better. There's another thing to think about. A lot of times they don't have standards for the two chemicals combined. You don't know if alcohol and benzol makes it worse. It might make it better.

### Joe, laboratory assistant, chemical plant

Somebody should take the data sheets that the manufacturers put out and explain them. The chemicals come into the plant with the original labels, mostly trade names. Unless you were a chemist or looked it up, you probably wouldn't know what it was. Most people don't know what the hell you're talking about with all these chemical terms. The bottom line is, "Do Not Inhale the Vapors," so they will turn their head while they're sucking it in.

### Tom, print machine operator, university

The chemical safety data sheets are very vague, and any employee who's going to assume his rights under the Right to Know law needs professional assistance. A competent chemist or environmental hygienist should sit down with him and explain the terms. I have a college education and I needed help understanding and identifying these things. Either it has to be put in language that people can understand, or there has to be more professional assistance. There should be an information handbook for employees defining hazards, such as occupational dermatitis, toxics, carcinogens, and outlining their rights in different work situations under state and federal law. Most people don't know they can say, "I might kill myself . . . I'm not doing that." Most people don't know they have that right to refuse. You have people using this stuff with bachelor's degrees and with an eighth-grade education. Both must be served.

## Conflicts among Experts

### Lee, stage carpenter, university

I haven't found anyone who knows what acetone does to people who inhale it. It has some properties similar to benzene, which is known to be carcinogenic. Whether or not acetone is harmful, I don't know, and I'm not sure how to find out. There are reference publications about hazardous chemicals available in the public library and indexes of industrial health hazards that are fairly current. But they're not books that you can pick up and read cover to cover. It would be like reading a dictionary. You also run into contradictory information. We had that problem with styrofoam and expandable polyurethane—getting reports that it's both safe and toxic, that it's both fireproof and highly flammable. In that case, we simply tested it and, sure enough, it burns like crazy. The information is there in some cases, not in others; reliable in some cases, not in others. It would be a full-time job for at least six months to become familiar with the current information about the things we use most of the time. That finds its place in my schedule way down the line, after all the plays are done and everything else is taken care of.

### Vivian, laboratory technician, research institute

The supervisor gives me information if he has it available. Sometimes he directs me to sources of information in the library, or to someone who has been working with it, but there's no guarantee that what's in print is current. There are always disagreements about the proper handling of chemicals. The manufacturers might recommend something different than my supervisor, who might feel it wasn't all that dangerous. To settle things we called NIOSH and OSHA to get safety data sheets with their threshold limit values.

### Mark, physicist, university

It's all a matter of who you believe. If the government says something is dangerous and there's some OSHA regulation—I may not always respect the people who made that decision as much as I respect one of my colleagues who says it isn't dangerous. In the scientific community, it's all a matter of whose opinion you think is best. And I don't have a very high opinion of the people who are doing research for the government. If you deal with chemicals a lot, you know pretty much from sight and from smell what's what. You don't wipe your finger over some chemicals that might dissolve it away. You don't mix your soup in an unlabeled beaker when there's something in it that's boiling and there's a hand reaching out.

### Bob, fire fighter

To my way of thinking, the state office building fire could have been settled pretty easily if someone would have just come in and told us a straight story about the dioxin. But one expert told us that if we got dioxin poisoning not to worry because there was no known cure. "You're going to die anyhow." Another told me, "You know, there's a possibility that this could have even been good for you, because it speeds up the process in your liver and it cleans your blood out." Yet another expert told us that because we were so close to the incident at the time, we had less chance of being contaminated than, say, somebody that was further away from the site. That didn't make sense. We were getting everything from you're going to die to it's good for you, everybody should have a dose! These were supposedly medical experts. So we're still in the dark as to what's what. One fellow says, "The reason you're getting these stories is because nobody really knows." I wish they'd tell me that. I'd rather have them tell me that they don't know than to have one tell me this and another tell me a 180-degree different story. Last spring a doctor at the local college who's supposed to be the foremost expert on PCB poisoning said that blood tests were useless because by the time we got them the dioxin was already lodged in the fatty tissue or in the organs of the body. Then a doctor from the Poison Center said that this "foremost expert" was not an M.D. so he wasn't qualified to tell us that. It was very confusing.

---

### Management Control of Information

Our respondents attributed their difficulty in obtaining information to deliberate management policy: "They keep it secret"; "They don't want to make waves." Workers felt that management kept control over information to avoid "mass hysteria," divert protest, and to prevent refusal to work. However, few were willing to challenge management's control over information because they were afraid to be labeled as troublemakers.

---

*James, computer assembler, manufacturing plant*

They're always going around with meters to see if exposure levels are within specs, but they always control the information. One area where I worked had a lot of dust. We knew it was bad because people were having trouble breathing. You had to blow your nose every half hour because your nasal passages would clog up. When I talked to my manager about it, he did bring someone to check it out. They brought a vacuum pump with a filter that would suck in the dust. But when it came time to tell us the results, they just said, "They're within company and OSHA specs." We could have fought for the information, but they would have stone-walled us. Then you go back to the fear of being labeled a troublemaker.

*Dick, granulator, pharmaceutical plant*

Right now the shop stewards can't go over to the medical division. If we go over there to argue anything, we can be suspended on the spot. We're not allowed to go there because about three years ago one of our stewards tried to get some information. She waited until the doctor got off and followed him to the parking lot, still trying to get her question answered. The doctor ended up calling the guard, and she ended up getting suspended for four days. Then they came out with this policy where no shop stewards could go to medical because they might try to argue.

### Sheila, laboratory technician, research institute

They worry that if you give people information, you're going to get mass hysteria. Well sometimes there may be an overreaction. I experienced it myself. You read about hydrogen sulfide and it's horrifying: coma, death, convulsions. I got upset when I read that because I never knew the concentrations. But any overreaction is compensated by lackadaisical attitudes. Probably 99 percent of the people won't read information, and 99 percent of those who do read it will say, "Yeah, that's really an extreme case."

### Walter, pipe fitter, glass factory

They don't tell you much about people's health. They keep it secret because the less we know the better off they are. They give some basic ideas and a bunch of numbers, but they don't really answer your questions. If I ask about the titanium tetrachloride I would expect him to tell me, "If you breathe it too long, it can cause severe lung injury." I know it plugs up the lines and smokes and eats the nailheads off the roof. When I was on the safety committee, I argued about stannic chloride. Everybody in the glass industry uses it because it keeps the bottles from being scratched. The company said, "It's water soluble. It won't hurt you." Now, that may be fine if you just smell the vapor one time and go away. But breathing this vapor constantly, day after day, in time it's got to take its toll. I've never come right out and asked, "How many people have been affected by stannic chloride?" Instead, I wrote to a professor at the local college. His reply was, "If you work with it, you're crazy." I showed that to them, and they admitted, "We've seen that before." However, without that hot-end coating they can't sell the bottles.

### Stuart, mold maker, glass factory

When a guy here got silicosis everybody was concerned, so we read a lot on it. What's silicosis? How do you get it? How did they find it? What's the doctor got to say about it? I don't want to depend on the company to supply the information I'm looking for. I want to know, What is silicosis, what is aluminum exposure, what is lead poisoning? What's behind it, what can happen to you? I'm sure that the last thing the company wants is for me to find out.

I just found out about the Right to Know law a year and a half late. I talked to a few guys in the shop who knew absolutely nothing about it. It was never posted. I made a request for the standards on everything they've got in the shop. The safety director said he'd get them to me, but when will he get around to doing it? I'm not so sure they are required to provide in-depth information. I want to read more than just the minimum requirements. Companies don't want you to know because they're afraid you won't come to work.

### Irv, plastics fabricator, aircraft factory

I would like to get notices properly posted in the plant specifying the dangers of some of these chemicals and solvents. The company didn't even post that Right to Know law because they were afraid people would bug them about information. They don't want to make waves. It's an irresponsible attitude, but they are engrossed in their business to the point where they don't see that people might get harmed.

### Jerry, materials handler, glass factory

I wonder what this stuff is going to do to me. That's why we're trying to find out what the material is. We went to our personnel manager and he gave us a real rough time. The company seems to have a real stick-in-the-mud attitude. I get the feeling that they don't want us to know. It's just a feeling I get because of the runaround we're getting when we ask for information. I don't know why. It's not like we're trying to shut down the plant or cause a big stink. We just want to know what we're working with and what it's going to do to us.

### Sally, services technician, hospital

I want to know how much ethylene oxide is really escaping from those autoclaves. I want to know how accurate those aerators are and why things aren't vented to the outside. I want somebody to find out what the actual levels are. We were tested about a few months ago because something went wrong and I'm pretty sure it was to keep us from suing the hospital. I think there should be some follow-up. There is something wrong, believe me. So I started reading this literature about ethylene oxide and how to monitor it. I took a brochure off the desk. They assumed that nobody up here is going to read the material and just left it around. But they forgot about me, old nosey. They're very upset about that. They would rather assume that I couldn't read or write because they don't want workers to read this type of thing. They don't want people to become alarmed. They don't want people to know. If you don't know, you stay here and get your work done.

---

### Alternative Sources of Information

Mistrusting official information, workers turn to a variety of other sources. Some told us they directly contact the manufacturer of the products they use for material safety data sheets. A few use reference materials in their local libraries. Others turn to co-workers or friends whom they perceive as knowledgeable about hazards. Some workers turn to COSH groups and to their international unions for technical advice. NIOSH too provides a substantial amount of information on health effects and control measures, much of it supplied by industry and channeled to local unions through workshops, conferences, and distribution of literature. University-based programs provide training and education through labor studies programs. Most of these efforts are directed to organized workers, leaving employees at nonunion firms to seek information as best they can.

### Kitty, industrial painter, university

I learned something about health and safety because I had friends who were on the health and safety committee for the union. You sit over a beer and talk. They opened me up to it: "Boy, those chemicals you use must be bad." I was invited to attend a union workshop on health and safety and picked up some things here and there. Now I drive 50 miles every Tuesday night to take a course. I don't believe there's anybody else in my shop that would take a 50-mile drive over and 50 back.

We're supposed to be able to take advantage of the Right to Know law but nobody in our shop has dared try it out yet. You have to go to the boss and make the request and nobody wants to because if you do he'll remember that you did and think you're a troublemaker. I'd like a decent union representative to come up and check on things, and I would like a bulletin board in the workplace that lists the damned chemicals on it so the guys can read it if they want. I'd like to have that provided for them so they don't have to take that scary little step of going to the boss. I'm not sure they'd read it, I'm not sure they'd pay any attention to it if they did read it. But it would be nice to have it there.

### Mike, photo lab processor, blueprint company

Some stuff is just downright ick, I mean, you just look at it cross-eyed and it attacks you, and you go cough, cough, gag, gag. It's common knowledge those things drive anybody crazy. But when we ask the company they don't tell us if it's dangerous. So we have to go to the manufacturers. They're obliged to say, "This is dangerous, take this precaution." A friend of mine has connections in the right places. She gets documents directly from the manufacturers and that's how I get informed.

### Ben, repairman, chemical plant

Sometimes I don't know what I'm working with. I take it for granted that if something is white, it's soda ash; if it's brownish white, it's lime dust; and if it's black, it's coal dust. So I can tell by looking. But as far as that goes, what is coal dust? What are the actual chemicals I work with? I don't really have that information. Most of the information I get is from my own research, which is mainly through the COSH group. If I need something, I call them up. I also go to the library for information and to the company safety office for pamphlets.

### Steve, railroad trackman

I went to high school and got little dribs and drabs about chemicals here and there, then I took a couple of courses that the union sponsored on safety. However, I'm an avid reader and amateur student of science. I take a very keen interest in science, which is beneficial to my role in the union, although it veers off into more obscure corners too. Formal education? Not that much, but in a sense I'm pretty well educated or, let's say, intellectual. Mostly I read newspapers and sections of magazines. I also read scientific journals and reports on current developments in science. I should read more books and texts.

### Stuart, mold maker, glass factory

My welder is having bad headaches. They found out that his blood pressure is elevated. Man, his lungs are all tore up too. He starts coughing and we tell him, "Why don't you get a gun, it's faster!" I have a book called *Work Is Dangerous to Your Health,* with a chapter on the hazards of welding. This guy's been welding 25 years. That's all he does, electric welding, acetylene welding, everyday in a confined area with no ventilation. If a dandelion poof ball flew in there it would float around all day before it got drawn out. You look on the floor and the powder is just solid, two or three inches deep. He's got an oven that heats the stuff up to 800 degrees. It has asbestos insulation. So he's really exposed. I showed him the book: "Look. It's real general. Nothing real heavy. Would you like to take a look at it?" He said okay and took the book. The next day I asked him what he thought. He said, "I don't know, might be a problem and might not be." So what am I going to do? If he's not going to press it, I'm not going to press it for him.

### Sue, laboratory technician, university

I'd like to see some sort of council that would be working specifically for the good of lab workers. It would be a scientific kind of thing. I wouldn't expect them to tell me if it's safe to work with this or that. They should just tell me the facts.

# 12  Controlling the Risks

When Alice Hamilton was seeking ways to protect the health of chemical workers in the 1920s, she often wondered about the insistence of employers on maintaining control over workplace hazards. She attributed this to "the survival in American industry of the feudalistic spirit, for democracy has never yet really penetrated into many of our greatest industries." Little has changed. Indeed, the question of control over health and safety in the workplace has become even more contentious than traditional disputes over wages, because it bears not only on the distribution of wealth but on the structure of the production process itself.

The issue of occupational health ultimately comes down to a set of political questions about industrial democracy, about power and control. What is acceptable industrial practice in the handling of toxic chemicals? Who determines acceptable risk? Who should control the conditions and the process of work? More broadly, who should control the process of technological change? Among the findings of the Michigan Quality of Employment survey was the widespread desire of workers to have input into decisions affecting their health. Workers who felt they did not have adequate authority and autonomy over working conditions reported significantly greater health problems.

Similarly, recent psychological research on stress links the incidence of heart disease to the extent of job control: those workers whose jobs involve psychological strain, but who have little ability to influence how their tasks are done, are more subject to illness. Low job control itself is a risk to health.

In our study, workers talk a lot about questions of control. Their concerns about control follow from their belief that working in a safe and healthful environment is not a privilege but a right and that this right is violated by management production goals. Those with little trust in management see the workplace as a combat zone over the right of workers to participate in production decisions in the interests of their health.

**The Right to Health**

Activists use the moral language of rights as an effective way to mobilize action and to instigate change. Workers talk about their problems of become informed about risk in moral terms. Sharing the widely held belief in the citizen's "right to know," they feel they have the right to information concerning risks on their jobs.[1] They also talk about their "rights" when considering the question of pay differential on the basis of risk. Most prefer not to take such risks, but, if they have to, they feel they have a right to bonus pay. This concept of "rights" pervades workers' perceptions; they talk about their right to work, their right to refuse dangerous work, their right to know, and, above all, their right to health.

---

*Ben, repairman, chemical plant*

It's not a privilege to work. It's my right to be able to work, to be able to go out and sell my services and be paid for it, to be able to feed my family, to be able to live and eat. I've got to have a job, I've got to be able to pay my rent. I've got to have an automobile to get to my job. I also have a right to know if I'm working with anything that's harmful. I should have the choice of whether or not to work with it. If I don't want to work with it, I don't feel I should be disciplined. There's nobody that has the right to tell me to do something that's harmful to me. I don't care what the job is. But if I lose my job, I can't just go out and get another one when there's 10 percent unemployment. If I got a job at $4 an hour, I couldn't afford my house, I couldn't afford my car, I couldn't afford to send my kids to school.

*Sandy, rigger, chemical plant*

We're paid quite well, compared to other workers. But I don't think that there's any amount of money you could pay a worker who will

1. The National Academy of Sciences Committee on Public Information in the Prevention of Occupational Cancer has argued that the "right to know" is "a fundamental ethical principle," following from "the individual's right of self determination." This "right" is also an extension of consumer values regarding the disclosure of product contents as reflected in laws regulating food and drugs, pesticides, and certain household goods.

contract asbestosis or lung cancer. I don't think any worker would accept a job if he were told that was the price. If a worker were to say, "Okay, I'm employed here and I know I'm going to get lung cancer," I would imagine the price of his services would be astronomical; the employer wouldn't be able to survive economically. Money is essential for survival on this earth, but it's self-defeating when you put your money above your health. You may achieve your financial goals but you won't live long enough to enjoy them. That's kind of contradicting yourself.

### Ken, electrician, chemical plant

We could force a company to give bonuses for dangerous work. If they did that, at least we'd go in with our eyes open. Now they pay a fellow who's not exposed to anything the same amount they pay the fellow who's exposed to everything terrible.

### Mike, photo lab processor, blueprint company

I'm just as human as management is, you know. I have to go out and get groceries everyday, pay for my house, pay for gas—I'm just as human. Why should they be exempt from the problems that I have? Why should they just lord over us all the time like we're animals? I just don't understand.

### Fred, chemical operator, chemical plant

I think we've got to give them a dollar's worth for a dollar. But I think we have a right to be treated equal. I'm a free man. I volunteered in World War II and I believed that we fought for freedom. Roosevelt told us we were fighting for the four freedoms and I believed what we were doing was right. I didn't come back to be subjugated by some company. You face the Japs and the Germans, you ain't gonna come back and be afraid of some pipsqueak company.

### Don, railroad conductor

Sometimes it seems that rights are so well-protected that it works against us. How many times do you hear about somebody who is guilty of a crime getting off on insanity? The company's rights are so protected that we can't go out and say, "Look what happened to my kid," because it can never be proven. Let's fight for *our* rights. It's actually easier to kill somebody in cold blood and get off on that charge for insanity than it is to prove that dioxin did anything to anyone.

### Steve, railroad trackman

Health and safety are connected to other issues. It's connected with your wages, because if you don't make enough money in 40 hours you have to put in overtime, and that's unsafe. It's related to your health and welfare benefits, because if you're constantly sick then your health and welfare fund isn't going to be able to buy much insurance. It's related to retirement, because if you die you won't retire and you won't collect any retirement. It's related to the right to strike, because if you don't have the right to strike, that is, the right to walk off the job and withhold your labor, then your boss is an absolute dictator who has the right to kill you because you don't have any choice.

---

### The Absence of Trust

Many workers perceive safe and healthy working conditions in terms of nonnegotiable rights, and assume there is a basic conflict of interest involved in respecting these rights. Thus, they tend to mistrust the paternalistic practices of management: "The company doesn't care." They also resent the laissez-faire policies of government: "They have no idea what goes on." In this context of mistrust, workers' control of health and safety appears as the only means to create conditions that would preserve their rights: "The responsibility has to be ours."

---

*Laura, filter cleaner, pharmaceutical plant*

A guy in the Safety Department said to me, "You don't understand toxicity. We'll take care of you, trust us, we're the ones who know, and we're the ones who are setting the standards. We want you to be healthy, so you should trust us." They put an analysis of the job hazards on the bulletin board in my area, and wherever it listed reactions like dizziness or vomiting, someone had written in "and cancer." We don't trust them. And we don't trust the government either. They deny that Agent Orange causes mutations, and to me it's pretty clear that it does. So I don't believe that the government works in my best interests. I don't think OSHA'd really steer you wrong, but it takes 10 years to get an answer from them. I'd be more apt to trust what comes from the international union. Those people are sincere and interested in health. They'll give me an evaluation that I can believe.

*Earl, landscape supervisor, botanical garden*

Trust is the assumption that some party has something in mind that's bigger than simple profiting off of you, that they consider your well-being. We know from experience that American industry does not consider people's well-being except insofar as the law requires it, and then only to the letter of the law. I have no trust beyond that.

### Lisa, laboratory technician, university

We need information from the unions about health issues. I don't really know why I trust them, except that I sort of believe that workers themselves are the ones who are going to be the most concerned about their own health. When I get information from my boss, I wonder why he's telling me these things and yet not complying with them. When I get it from the university, I feel like this probably doesn't apply to me. It's just being sent to every person who works here. Usually you can't understand what they're talking about anyhow, so I read it quick and put it in the drawer.

### Ted, welder, chemical plant

We didn't know we were working with nine recognized carcinogens until we had a health and safety study done by an industrial hygienist. This is the company, these are the people who are supposed to be taking care of you. God, these are the people who told me this stuff isn't going to hurt me. They said hello to me in the morning, "Hey, how're you doing? How's the family?" They never told me that I was working with carcinogens. That really set me off.

### Ken, electrician, chemical plant

For years they told us that if cyanide doesn't kill you right away, it won't hurt you at all. They never told us it has a cumulative effect. These are engineers, these are men who have quite a bit of learning, so they must have known different. They couldn't get anybody to work for them if they had told us the truth. We were accumulating this in our system and it was hurting us. It's hurting us now. I do trust people, but I'm beginning to find out. . . . It sort of shakes your trust to hear a man who has a doctorate degree in chemical engineering tell us that cyanide doesn't hurt you. It's almost like the asbestos case. They knew before Christ that asbestos was bad for you. We're talking about 2000 years ago. I do trust people, but people just worry about one thing, themselves, their position.

### Rose, pill coater, pharmaceutical plant

Some people don't want to think about risk. They believe, "If it was really going to hurt me, this company wouldn't let me work on it." But the company doesn't care. People are going to have to be more educated about what they work on. Chemicals are hazardous, but they don't want to believe that.

### Bill, fire fighter

I don't trust politicians to make decisions about something that's going to affect me. They're not fire fighters and they've never been inside a burning building. So they have no idea what goes on. After all, they wouldn't trust my decision on a governmental policy.

### James, computer assembler, manufacturing plant

You just can't trust the company having all that information about our medical history without workers having their own experts. We've seen too many situations where the company doctor or medical department has hidden the hazards for years. They're all the same as far as I'm concerned, and they're only trying to protect their butts from a lawsuit or compensation. Money is the name of the game. It's unfortunate that it has to be like that. As long as we don't have our own experts, we have to rely on theirs, and I don't trust them.

### Joe, laboratory assistant, chemical plant

I wouldn't trust that son-of-a-bitch company doctor as far as I could throw a car.

### Mary, housewife, wife of a railroad conductor

I don't want to vote in an election. For what, for who, for what change? There won't be no change. There will never be legislation that will protect us the way we should be. We have a responsibility to teach our children that, just because someone told you something is safe, it's not. Just because the federal government says something is okay, doesn't make it great. I don't think they have our best interests at heart. I don't trust them. The responsibility has to be ours.

---

### "Controlling Our Lives"

Democratizing the workplace has been a persistent theme in industrial reform movements, especially in western Europe. There have been relatively few worker-control experiments in the United States. However, during the late 1960s and early 1970s experiments on worker control, often borrowed from Scandinavian models, proliferated in response to spreading absenteeism, growing alienation among workers, and declining productivity. Visions of worker control, however, differed depending on their source. For the advocate of industrial democracy, worker control meant the power of workers to control decisions about the conditions of work and the processes of production, not just through the wording of contracts but on a day-to-day basis. For the managers, experiments in worker control were intended to increase participation in decisions within a carefully circumscribed arena. To the extent that they encouraged worker participation, they hoped to improve morale and increase productivity; participation was a way to increase worker commitment to managerial goals.

Attitudes among workers themselves reflect these different visions. Activists believe that workers must have real power over their work environment, both through their right to refuse work without retribution and through their collective voice in decisions about the process and conditions of work. However, the vision of most workers is more limited. Immersed in the day-to-day reality of work, they talk of control less in terms of assuming power or "taking over" than of contributing to the resolution of specific problems.

While people demand greater control, they are vague about how to exercise such control. Some talk about collectively withholding labor. Others hope that stronger unions will resolve their problems. Several workers talk about joint labor-management forums and consensus decision-making models. The theme of control runs through their discourse on work, technology, and politics.

---

### The Workplace

*Ben, repairman, chemical plant*

I'd really like to be in some kind of joint labor-management organization, where everyone sits down and forgets about who's boss and who's paying who, and thinks about what we can do to solve the problems. What can we do to keep the productivity up? What can we do to keep the economy strong? What can we do that's best for both our employees and employers? Can't we combine them together?

*Dick, granulator, pharmaceutical plant*

It's management's obligation to police its own plant to ensure a safe and healthy workplace. But workers themselves are the best qualified to recognize problems because they're in the job everyday. We like to think of ourselves as the eyes and ears of the company, seeing what they don't want to see. And the mouth. This year in our contract we're going for stronger health and safety language and more freedom to have more say. They should work with us, instead of always trying to be big daddies saying, "We'll take care of it. Be a good boy and don't make any trouble." The way it is right now, it's pretty one-sided. We're supposed to be a joint management-union safety committee, but it seems like the union is just bringing up the questions while management has the say.

### Irv, plastics fabricator, aircraft factory

We've tried several times to get management to recognize the joint management-union safety committee as a viable force. But because it's a family business they find it very difficult to give any authority to anyone outside. They are paranoid about letting their power go. They advertise the safety of their sailplane. I wish we could get it into their heads that it's not just the sailplanes they have to worry about but the people that build the planes as well.

### Walter, pipe fitter, glass factory

They're starting a program called the "quality circles." Through the quality circles, we're hoping they'll start listening to the people who are the experts. When you do a job day in and day out, you're the expert on the job. You know how it runs. Management can design the machine, but the man that runs it for 10 years is going to know more about that machine than management. It's that simple, and they're starting to realize it. I'm hoping for a little change.

I've spent eight months now setting up these quality circles with the seven management people and seven unionists on the steering committee. When we first started meeting, we were maybe a foot apart, looking at each other with no trust. Right now we're intertwined like the fingers on your hands. I say what I think. The management guy says, "I may not agree with you but maybe you've got a point." They bring out their side and we bring out our side and we've come up with a lot of good ideas. There's no vote so every decision is made by consensus.

I like it when we meet with management people because you get both sides of the story. The more you learn about both sides the better off you are. The bottom line is, if the company you work for doesn't make a profit, you don't have a job. If they do make a profit, in one way or another—benefits, vacation, holiday, whatever—I'm going to get my share.

### Stuart, mold maker, glass factory

You know, this machine's got a memory bank, that machine is semicomputerized, and they're going to bring in these machines that are 100 percent computerized. I'll tell you something—what's to stop them from running a conduit full of wires into some office in there and have management control all your feeds and speeds?

### James, computer assembler, manufacturing plant

If the company keeps de-skilling jobs, keeps sending the good jobs out, keeps bringing in more and more chemicals, which I know damn well are going to cause problems, I think we might have a good shot at an organizing campaign. It could open up a whole new way for us. We could get shot down bad, but we have to do it. With more and more people being laid off in this town, it leaves us as privileged workers, and I don't feel we should be bought off. As workers, we should have a certain solidarity with our community and our class. The company tries to break up unity among workers by keeping everything secret. They keep information away from the community on the groundwater problem, they keep it away from us on what we're working with in the plant. They're not giving us test results on air or noise samples. They try to make us think it's too difficult to identify hazards and to understand them. It's not that hard. All they're trying to do is snowball us into thinking that it can only be done by their experts. We're being treated like second-class citizens. We have no choice but to organize just to save ourselves.

### Steve, railroad trackman

Enforcing safety is a struggle, tooth and nail, with management. There are those who say that labor and management share a common interest in safety, that safety is productive and cost-effective, that it saves management money on claims and lost working hours, that all we have to do is get together to work on it. Well, management is not exactly stupid, and if it's true that safety benefits management, why do we have to fight them so hard? Obviously there's some fault in the reasoning. Safety is much more a point of contention—us versus them—than a field of common interest. Not that they don't have an interest in safety and not that they won't sometimes enforce safety regulations, but, by and large, if the union's involved, they don't want to have anything to do with it. It's a daily struggle, us versus them. They try to get as much work out of us as they can, and pay the least possible attention to safety and working conditions. We try to live through it for 25 or 30 years so we can retire.

### Nick, chemical operator, chemical plant

We have a lot more power than workers realize, but we don't use it to its fullest extent. People have to say, "No. Until you fix that, I'm not going to run it." The power is there, but we are still not willing to stick our necks out. It's a big problem to get our guys to say "No." They are always looking for somebody else to tell them, "Don't do this," so if there's some kind of discipline, they can always say, "Well, he told me not to do it." They aren't willing to take it on themselves. I'd like to see it come to a point where, if an operator says something is wrong, then the equipment—right then—is changed or fixed or shut down until it's right. But until there's more power coming down along the line, you aren't going to see that.

### Arnie, chemical operator, food processing plant

You know, you learn about the civil processes of industrial and labor relations, but it's all hogwash. Labor relations are based on one thing, and that's power. And information or knowledge only has value to the extent that it plays into power. Information isn't an end in and of itself. It's only useful as it helps support the power of one group or another.

In the sixties and early seventies it seemed that our whole society was coming around. People were relying more on militancy, on unity and solidarity, than they had for decades. Then, in the late seventies, the movements were sucked into these legalistic channels of resolving disputes controlled by the ruling class. Supposedly in this country we live in a democracy, which is a crock of shit. We have some forms of political democracy but no form of economic democracy. The people who work and the people who are affected by those "eco-systems" in the plants should own those "eco-systems" and have a voice on how they are run. Now that by itself won't necessarily bring about a better society. Just because 30 people own a plant doesn't mean they'll be any more compassionate, especially with the values that we have in our country like "any means justifies an end." The whole fabric of the country has got to change in order for such economic changes to bring progress.

### Lee, stage carpenter, university

As far as protection of health and safety on the job is concerned, the only adequate safeguard is to have the people who are exposed to hazards, the workers themselves, be substantially in control of their circumstances, their physical surroundings, and the priorities involved in their work. My situation here is much better because I control what goes on. I feel much safer now than I did in previous jobs. There were fewer hazards and they were less severe, but I had absolutely no control over what I was doing, when I would do it, or what rate I would do it at. In those cases, I would feel powerless, and that directly relates to the danger of being injured.

### Laura, filter cleaner, pharmaceutical plant

I'd like to get rid of that awful control hanging over us. We have to have more control of the workplace, the environment, and the people who are running the government. I would feel better if I knew that I could go into work and say, "There's no way you're going to do this, that's it, forget it. I have information that says that it's wrong, so let's just do it this way and not get ourselves into a situation where someone gets hurt." But they have us by the balls.

### Earl, landscape supervisor, botanical garden

I can't say the work is as safe as I'd like it, but we've been working on this as a group for about a year. We spent a great deal of time in the winter involving everybody at the gardens in planning a pesticides policy. We still haven't gotten fully developed procedures for all kinds of things, but we've got a much more acceptable system. So in some respects we're a lot safer than before.

### Don, railroad conductor

Maybe we have to take things into our own hands. If there were only a way that the people directly affected could decide what is used on the job. Ultimately we're the ones who are going to be exposed, not the supervisor or the superintendent who's sitting up on the third floor in his office somewhere. It's the union members, the train personnel, or the nonoperating personnel, the track workers, the signalmen. They should be the ones ultimately who decide their own fate.

## Technological Change

### Steve, railroad trackman

Technology can mean a lot of things. Technology for the bosses means automating people out of jobs, or maybe spraying them out of jobs. By using toxic chemicals they can avoid using maybe ten times the number of employees with clippers. There's nothing inevitable about that. But technology can also be designed to benefit the people. It could mean coming up with various "natural" or "organic" means of doing what they now do with chemicals. That could eliminate introducing toxins into the environment. So, it's the abuse and misuse of technology that I'm against, not technology itself.

### Arnie, chemical operator, food processing plant

In America we separate the production process from the social process. We view progress according to the efficiency of production, not its social impact. I don't think technology necessarily means progress at all. There are so many examples of where it hasn't brought progress, the kepone incident[2] being an obvious one. How can we say to the people who worked at that small Allied subsidiary that there is progress in this country? There was no more fuckin' progress for those people than there was for people who had to beat their brains out in a foundry in the late 1800s, and probably died at 50 or suffered neurological damage because of the things they were exposed to at work. For those people there was little progress in terms of social justice. It's the use and control of technology that ultimately defines progress in society.

### Les, furniture restorer, self-employed

I don't like the dependence that seems to come with new technology. I won't wear a digital readout watch, for instance. I want my watch to have hands, not numbers. I don't like that stuff; it's out of my control.

### Ken, electrician, chemical plant

A lot of things that are good for me and give me enjoyment come out of all this technology, things I wouldn't want to give up. It's nice to return to basics, but I don't know if I would want to live on a farm and cut my own wood. I know that things that give me pleasure are also bad for me or somebody else. This chemical company probably makes a lot of what I use. If you shut the company down, it wouldn't bother me, but it probably would in a roundabout way. You can't shut down every company. The chemical industry in a sense might be like the atomic industry. I think they got locked into atomic energy and don't know how to get out of it. We are locked into the chemical technology and my sons are going to have to pay the piper.

### Don, railroad conductor

Everyday it's something new. Always something new on the market; a new drug, a new chemical, a new procedure in medicine. It's scary; we enjoy these new techniques today, but what happens five or 10 years down the road? No one looks at that. It's always, develop a technology now, and if something goes wrong, we'll take care of that later. But the increased use of chemicals in the past two decades has caused a lot of cancers around. In my parents' generation, the incidence of cancer was very low. People died of heart attacks, of old age, but rarely of cancer. Even in backward countries the cancer incidence is low. It's mainly a problem in your big industrialized countries where they have all these new chemicals. Hopefully, when my kids grow up, companies will have

2. The kepone incident was a case of severe overexposure to the chemical in a factory in Virginia, where at least 70 employees and 10 family members suffered acute chemical poisoning.

to do a lot more studies before they just pop a new chemical on the market.

### Ted, welder, chemical plant

I've got a 1971 Ford pickup outside that I paid $3100 for. If I bought the same truck today, 11 years later, it would be a piece of crap. All this high technology hasn't helped me or my family. It's polluted the air and the water. While it's given me cheaper products to buy, they are more toxic. Technology is designed more to put profit into the company's pocket than to make a better product for the consumer or to give us a safer, better life.

### Jerry, materials handler, glass factory

Technology is great as far as it goes. It makes people's lives easier. It gives you more pleasure time. Ten years ago, how many people had a microwave oven? Now a lot of homes have them. Bam, bam, bam, and shoo—you have a dinner in 10 minutes and you're off to the races.

But technology can also put people out of work, or make them sick. I was reading an article about workers having children with birth defects and about the sterility among younger people. Technology has a tendency to cause its own problems. Until you can control the health problems from the advance of technology, there's going to be trouble.

### Gene, pipe fitter, chemical plant

The company has its own treatment plant for some wastes and others they ship out. There's a facility supposedly designed for that. They fenced in 50,000 acres of government property in New York State. We were shipping a lot of bad stuff up there, and they just dumped it into big open pits. All they did was move the problem from one place to another. It was legal because it was government sanctioned. Local people don't appreciate it too much. I've got family in that area. They wrote western New York off with Love Canal. It's funny—they publicized Love Canal and 10 miles away is a place called Bloody Run, which is 10 times worse. If you go another five miles, there's two big container tanks storing radioactive waste. You go the other way about 15 miles, and half the schools are built on radioactive waste. They're just trying to forget that end of the State. They opened a big can of worms with Love Canal, but that was only the top of it. It's a free society to a point. It scares you when you really think about it. You can't live without chemicals, but if they don't open their eyes and start looking at the long-term effects some bad prophecies will be fulfilled.

### Political Concerns

*Mike, photo lab processer, blueprint company*

This country is in a mass of problems right now. That's common knowledge. You take any situation, if somebody's profiting somebody else is losing. If you really look worldwide—everybody's hurting. Russia's in trouble, we're in trouble, China's in trouble. Everybody's in trouble. There are no winners, everybody's pssch!

There's coming a time when there's going to be a worldwide money system. There's this master computer with this scanning system in Brussels or Belgium or some such place. They'll implant on your hand or forehead a number by a laser system and this scanning system will be used in this worldwide bank. The computer is already working. The whole world is hooked up to it already. You'd be surprised just what is going on. Now, everybody's economy is suffering, and what's gonna happen? You've heard of electronic funds transfer? That means money is already on its way out. This country is already shot. The world is ready to crumble.

*Mary, housewife, wife of a railroad conductor*

We have to get mad enough to do something violent. That's why I like the way some Vietnam veterans are handling their claims. The average person doesn't want to hear it. But let them be affected and they'll understand why we get so angry. It will take an awful lot of angry people, people like us who've lost a kid. It will take a lot of money, a lot of time, and a lot of clout to make things safe.

*Laura, filter cleaner, pharmaceutical plant*

I believe in a democratic society where people control their government. You have to elect people that say they're for you, and then you have to make them do things for you. Unfortunately, a lot of people feel that their vote doesn't count. But in fact you have control over their jobs if you just don't reelect them. You have the authority to get rid of them. People have leverage if they would only use it.

*Walter, pipe fitter, glass factory*

The government is too big. But they're starting at the wrong place to cut it. Where do they always start to cut it? At the bottom where the poor man lives. The fat man that's making $100,000 a year, they never touch him. The rich get richer and the poor get poorer and have to work harder. That's what it amounts to, and people are realizing this. Sometimes I think the politicians think we're a bunch of boobs. We aren't all that dumb. We may be common people. We don't use big words. But we know what's happening. We're being taxed to death. We're paying into social security and getting it shoved right down our throat by a higher retirement age because some congressman, judge, or lawyer wants to work until he's 90. Hey, if I could only go sit behind a desk or go to work just when I want to go to work. . . . More and more I feel the politicians aren't listening to the people. When they do listen, they only listen to what they want to hear.

### Ben, repairman, chemical plant

I believe that the average middle-class person needs a voice. Most middle-class people think they're rich, but they don't realize that if they lose their job and their paycheck for a couple of months they're no longer middle class. We can lose everything we own. So we've got to have a voice in government to make sure we don't lose what we've got. Big business will eventually destroy our country if they continue the way they're going. Rather than fix a health problem in the plant, they would just as soon go overseas and build another plant. They think that's justified. Is that the answer? I don't think it is. They're only destroying our country.

### Lee, stage carpenter, university

Politics to my way of thinking means that people recognize each other and then begin to find some way of working together. I don't think that means Democrats and Republicans getting together on a bipartisan tax bill. There's no widespread recognition that there are some people whose interests are opposed to industry. There's no widespread recognition that public policy has been used against the majority of working people. There's no widespread recognition that progress has removed people's control over their own jobs. Until these problems are recognized, there's not going to be anything in the political arena that looks like politics to me. I work in the theater, and I know illusion when I see it. What we've got going on now is the rerun of a bad show.

# Conclusion
## The Social Dimensions of Risk

Through their direct experience workers have offered a variety of insights into the problems of working with chemicals, the sources of these problems, and possible solutions. Looking for common threads running through the interviews, we were first struck by the diversity in their perceptions: some saw risks as "dangers," others as "part and parcel of the job"; some expressed anxiety, others the satisfactions that made the risks worthwhile; and some resigned themselves to working in hazardous conditions while others sought to change their working environment. Yet as they related their experiences, we began to see certain patterns in the way social relationships, feelings about work, choices, and, above all, their control over working conditions were tending to shape their perceptions and guide their responses to workplace risks.

### Social Relationships
When workers talked about their relationships at work, they often conveyed a pervasive sense of psychological and physical isolation from co-workers, management, government, and even their families. The persistent anxiety: "It's always on the back of my mind"; their reluctance to talk about health: "It's just nothing to talk about. Who would listen?"; and their fear that complaining about hazards could jeopardize their job: "I wouldn't

ever call OSHA. I mean, it's a good way to make your job impossible";
all fostered this sense of isolation. So too did the attitudes of co-workers
and families. Chemical operators sometimes found that even their friends
were reluctant to visit them because of odors or potential dangers: "People
don't come down here because of the smell. Can't blame them, really."
They faced complaints at home: "I walk into my house and they tell me
I stink."

Compounding their feelings of psychological isolation was their
physical separation from each other when using protective equipment.
The worker in a "bunny suit" and respirator is insulated from communication as well as from hazards; the "protected" worker is in a "clamshell" or "cocoon." Where prevailing norms discouraged the use of such
equipment, those who wore it felt ostracized or strange, "like a masked
marvel."

Management safety policies that rotated workers in hazardous
jobs created a pattern of segregation by breaking up work groups and
preventing discussion about common problems. Many maintenance workers, railroad workers, artists, and others who typically had little interaction with their co-workers tended to believe that their problems were
unique. And in some cases, the pejorative connotations of problems such
as sterility, cancer, or nervous disorders made people reluctant to talk
about their health. Thus we found that many people dismissed their problems as personal, failing to recognize that their co-workers were troubled
as well.

The limited means of recourse further reinforced social isolation.
Workers felt frustrated by company doctors who dismissed their complaints, and with supervisors whose interest in production blinded them
to potential hazards. They found that their unions, with limited resources
to deal with occupational hazards, often concentrated on other priorities.
And looking beyond their firm, they find their access to OSHA increasingly difficult. Thus, when employers, unions, and government inspectors
failed to deal with complaints, or rejected them as based on "unscientific"
evidence, workers inevitably began to doubt their own perceptions: "We
tend to blame ourselves."

### Attitudes toward Work

The perceptions of risk among the people we talked to were
closely related to their feelings about their work. Here the contrast between professions or craft workers on the one hand and maintenance or
production workers on the other was sharp. The beautician loved hair
styling, the fire fighter found satisfaction in saving lives, the sculptor
enjoyed creating an original work of art. Enjoying their work and valuing
its results, they tended to minimize the significance of its risks: "I love

what I do, so I don't really feel that I'm taking that many risks." In contrast, few of the blue-collar workers we interviewed expressed this sense of job satisfaction. While professionals told us, "It's worth the risk," production workers said, "We have no choice." We found similar variation among laboratory workers. Their perception of risk depended less on the nature of their day-to-day work than on the extent to which they identified with the value of their research.

The relationship between work satisfaction and risks was most evident when people talked about their children. A fireman was proud of his son who had joined the force. However a chemical operator faced with work which seems to offer few rewards for its risks hoped for "a better life for my kids."

Attitudes about supervisors and managers also helped to shape risk perceptions. Where mistrust prevailed, people tended to place the blame for workplace hazards on management practices. Factory workers told us that production took priority over health, while some laboratory technicians felt that research took priority over people. Some blamed health problems on inappropriate supervisory style, such as the absence of role models or appropriate guidelines, and the distance between management and actual operations on the shop floor: "Things don't get done until somebody gets hurt." Callous supervisors became a source of cynicism: "They say hello to me in the morning—'Hey, how's the family?' They never told me I was working with carcinogens." Workers felt they were being used as "guinea pigs." They described their feelings of "betrayal," expressing their cynicism in black humor: "If you're working with a piece of equipment, be careful. You we can replace. That we can't."

While work satisfaction colored risk perception, we also found evidence that attitudes about risk affected employee morale. Many people were intensely aware of the inequities in their situation: "I think we have a right to be treated equal." "I'm just as human as management is. Why should they be exempt from the problems I have?"

### Choices

For most workers, a job is primarily a source of income. Thus, workers talked about the risks of their jobs in the context of their employment alternatives. Burdened by family obligations in an uncertain economic and employment climate, few were willing to risk changing jobs, or even risk being fired for speaking out about workplace conditions. Those without opportunities to find alternative employment saw themselves as forced to choose their job over their health: "You can never balance the wage against the risk; you balance the wage against the alternative. The alternative is starving."

The ability to choose safer employment affected their evaluation of and response to risk. Workers talked a lot about the realities of supporting their families and paying their bills: "We're really trapped." Those with few choices were most afraid to complain. They felt they could not afford to be labeled troublemakers. Thus they lapsed into attitudes of resigned compliance and passive adaptation to hazardous conditions: "What you can't change, you accept, so we kid about it. Cyanide is good for your health." Those who reacted to hazards with more active efforts to change working conditions tended to have fewer economic constraints, greater employment opportunities, or a union that provided protection and support.

### Control

With limited alternatives, workers must come to grips with their existing work environment. Thus many were preoccupied with questions of control. They expressed their sense of impotence in the face of possibly irreversible and uncontrollable harm: "It's that time bomb that might be ticking inside you." The uncertainties about their exposure and the long-term effects on their health—"whether I'm going to make it to the end"— were a special source of anxiety and distress, especially for younger workers. Contributing to their sense of impotence was the technical complexity of the information they received: "There's a word in [the precaution sheet] like trichloroethylene. I wouldn't even know how to pronounce it, let alone understand it." The inability to use such information compounded concern about control: "The company will give us information . . . but it's like handing a Stone Age man a rubber grip for his club. . . . What the hell do I do with it?"

Workers' sense of powerlessness in part reflected their lack of confidence in supervisory and management efforts to control workplace hazards. We heard complaints that managers poorly understood the realities of the shop floor, yet ignored the judgments of workers who were in a position to offer solid contributions. Resentful of a hierarchical system that discounted the validity of their experience, workers referred pejoratively to "those educated men" or to "those men behind the desk." "They think we're a bunch of dummies because we don't have a degree." Those in a position to exercise their judgment about workplace conditions with some autonomy—the dry cleaner, the gardener, some laboratory technicians—were generally less worried about risk.

Most of our respondents viewed the issue of occupational risk in an adversarial context. They saw the desire for profits conflicting with the cost of protecting their health. Consistently referring to their companies as "they" or "it," they suggested again and again that economic interests conflicted with concerns about health: "They put up a facade of being safety conscious, but the reality of working conditions is of little concern because fixing things would reduce profits." In this context, they mistrusted the paternalistic practices of management and the "objective" policies of government. Few were inclined to talk about risks as something to be objectively measured and balanced. Rather, they talked in personal and moral terms about dangers that should be avoided at any cost: "The government, those people who set standards, they say so many parts per million and it's perfectly safe. I can't picture any exposure to phosgene or cyanide as safe."

The issue of occupational health is increasingly volatile, raising questions of distributive justice in the most fundamental terms. Inequities are clear. Production and maintenance workers are more likely to be exposed to chemical risks than white-collar or professional workers, and they tend to have fewer employment choices. The problem of occupational risks is one more factor in the widening gap between such groups, encouraging resentment and a fear among many workers that society considers them a marginal and expendable class. Many workers felt that management "couldn't care less if you died."

One solution that is often proposed to minimize inequities is to give workers supplementary wages for taking risks. Some workers expressed their willingness to accept the concept of hazard pay, if there were no alternatives. However, our interviews suggested that such a policy would only increase their burden of coping with the impossible choice between health and work.

Other proposals, focusing on the problems of technical expertise in regulatory agencies, are for science panels or hazard assessment groups to assist in evaluating workplace risk. While this could enhance the competence of regulatory decisions, our interviews indicated that most workers would not accept risks simply on the basis of managerial or governmental judgments, even when these are grounded in expert risk-benefit calculations. With their health and their family's health at stake, they want a say. They want to know what risks confront them in the workplace; they want a greater voice in decisions about the production and use of products that may affect their health.

This desire to participate was frequently expressed, but tempered by feelings of dependence. In the face of limited alternatives, workers were ambivalent about the best way to deal with risks. Those who felt their alternatives were limited aligned themselves with the interests of their employers in spite of their fear of risks: "They're trying to make a profit, so they let things go. The bottom line is, if they don't make a profit, I don't have a job." Others saw labor and management as sharing an interest in health as well as profits. They looked to increased participation of workers in cooperation with management as the route to controlling workplace risks: "I'd really like to be in some kind of joint labor-management organization where everyone sits down and forgets about who's paying who, and thinks about what we can do that's best for both our employees and employers." Finally, those with the most liberal opportunities and strongest social support looked explicitly toward an expansion of worker control. As one activist put it, "The only adequate safeguard is to have the people who are exposed to hazards in control of their circumstances. My situation here is better because I control what goes on."

Hearing these voices, we believe they carry a critical message—that the pervasive presence of chemical risks in the workplace has profound human costs in terms of anxiety as well as of illness. With the proliferation of chemicals in so many occupations, such concerns are likely to have an increasing effect on collective bargaining, on compensation claims, and on the general morale of the work force. Thus the voices of workers, their identification of problems, their insights, and their views must be heard. They are critical to the creation of a more humane working environment.

# Appendix 1
## Biographical Information and Index[1]

| Name | Biography | Page References |
|------|-----------|-----------------|
| Ann | Silk-screen supervisor, museum. Age 33, single, no children. 13 years at museum. Union member. | 88, 129 |
| Arnie | Chemical operator, food processing plant. Age 30, single, no children. 4 years at plant. Currently works as health and safety specialist for a union. COSH group organizer. | 7, 91, 107, 109, 120, 139, 155, 172, 173 |
| Art | Laboratory technician, university. Age 59, single, one grown child. 30 years in various laboratories at university. No union. | 86, 95 |
| Ben | Repairman, chemical plant. Age 30, married, 2 children. 5½ years as carpenter and repairman at this plant; previously held various carpentry jobs. Political action chairman, union local. COSH group member. | 12, 31, 72, 92, 104, 107, 117, 134, 143, 154, 161, 164, 169, 177 |
| Bess | Diffusion analyst, manufacturing plant. Age 52, married, 3 grown children. 3 years in job; 30 years at plant. Union member. COSH group member. | 34, 55, 85 |
| Bill | Fire fighter. Age 34, married, 2 children. 8 years in job. Father, also a fire fighter, died in work-related accident. Vice-president, union local. | 18, 41, 98, 119, 141, 168 |

1. This table includes only the 65 respondents from whose interviews we excerpted the direct quotations in the text.

Bob  Fire fighter.          43, 45, 61,
     Age 51, married, 6 children. 28 years in job; brothers 74, 97, 141,
     and 1 son are also fire fighters. District vice-presi- 157
     dent of union.

Carol  Laboratory technician, university.       77, 78
     Age 32, single, no children. 2 years at university;
     7 years total in various research laboratories.
     No union.

Daniel  Chemical operator, chemical plant.     53, 85, 88
     Age 24, single, no children. 6 months in job. Varied
     previous employment. Union member.

David  Chemical operator, pharmaceutical plant.   101, 119
     Age 24, married, no children. 5 years at plant; wife
     works there. Shop steward of union local.

Debbie  Hair stylist, beauty salon.        17, 27, 77
     Age 29, single, 1 child. 11 years as stylist; also
     manages beauty salon operation. No union.

Dick  Granulator, pharmaceutical plant.     48, 102,
     Age 25, married, no children. 3½ years in job, 6 132, 140
     years total at plant. Shop steward and health and
     safety committee member of union local. COSH
     group member.

Don  Railroad conductor.         49, 92, 128,
     Age 37, married to Mary, 4 children, 1 died of can- 144, 154,
     cer attributed to his work exposure. 12 years in job. 165, 173,
     Former union member. Currently holds management 174
     position.

Dorothy Deckhand.            9, 26, 77,
     Age 26, single, no children. 3 years as deck- and 84, 94, 99,
     dockhand on variety of ships. Union member. 119, 127

Earl  Landscape supervisor, botanical garden.   44, 49, 72,
     Age 39, married, no children. 1½ years in job, su- 75, 128,
     pervises 10–20 people. Worked in gardening 12 years 151, 166
     total, previously taught high school. No union. 173

Ed  Maintenance mechanic, university.     106
     Age 59, married, 1 grown son. 7 years in job, previ-
     ously steamfitter and building engineer, 35 years
     total at university. Shop steward and health and
     safety committee member of union local.

Elise  Laboratory technician, research institute.   43, 73
     Age 31, single, no children. 5 years in job, previ-
     ously in another research laboratory. No union.

| | | |
|---|---|---|
| Ellen | Laboratory technician, chemical plant. <br> Age 32, single, no children. 13 years in job. Executive board member of union local. | 40 |
| Eric | Sculptor, self-employed. <br> Age 31, single, no children. 14 years as sculptor; 4 years at museum, currently works on commission. | 16, 26, 43, <br> 76, 97, 153 |
| Eve | Sorter, manufacturing plant. <br> Age 44, single, 4 children. 17 years in plant. Previously worked 10 years in shoe factory. Shop chairman of union local. COSH group member. | 26, 62, 96, <br> 104, 119, <br> 134, 142 |
| Frank | Chemical operator, chemical plant. <br> Age 29, married, 2 children. 7 years in job. Previously worked 4 years at cosmetics factory. Shop steward of union local. | 88 |
| Fred | Chemical operator, chemical plant. <br> Age 62, married, 2 grown children. 34 years in job. Shop steward of union local. | 5, 27, 80, <br> 87, 115, 165 |
| Gene | Pipe fitter, chemical plant. <br> Age 40, married, 2 children. 8 years in job, 18 years total at plant. Former official of union local. | 79, 96, 102, <br> 107, 122, <br> 145, 175 |
| Greg | Air-conditioning repairman, university. <br> Age 32, married, no children. 1½ years in job. Previous employment varied. Union member. | 39, 55, 114, <br> 131, 139 |
| Henry | Rosarian, botanical garden. <br> Age 28, single, no children. 2 years in job, previous employment in gardening. No union. | 77, 99 |
| Irv | Plastics fabricator, aircraft factory. <br> Age 38, married, 3 children. 18 years at plant. President of union local. | 73, 86, 99, <br> 116, 160, <br> 170 |
| Jack | Nurse, hospital. <br> Age 27, single, no children. 2 years as R.N.; previous employment in health care. No union. | 53 |
| James | Computer assembler, manufacturing plant. <br> Age 32, married, 1 child. 1 year in job, 9 years total at plant. No union; union organizer. | 11, 35, 38, <br> 57, 59, 65, <br> 78, 79, 82, <br> 88, 103, 106, <br> 108, 120, 131, <br> 141, 158, 168, <br> 171 |
| Jenny | Laboratory technician, pharmaceutical plant. <br> Age 36, married, 2 children. 3 years in job; previously full-time house wife. Shop steward of union local. | 58, 120, 143 |

| Jerry | Materials handler, glass factory. Age 26, married, 2 children. 7 years in job. Shop steward of union local. | 108, 160, 175 |

Jerry — Materials handler, glass factory.
Age 26, married, 2 children. 7 years in job. Shop steward of union local. — 108, 160, 175

Jill — Dialysis technician, health clinic.
Age 28, single, no children. 3 months in job; 5 years total at clinic. Executive board member of union local. — 15, 74, 152

Jocelyn — Secretary, museum.
Age 32, single, no children. 1 year in job; previous employment varied. No union; union organizer. — 60

Joe — Laboratory assistant, chemical plant.
Age 37, married, 2 children. 4 years in job; 19 years at plant. Executive board member of union local. — 36, 60

John — Maintenance worker, food processing plant.
Age 27, single, no children. 3 months in job; previous employment varied. No union. — 94, 130

Ken — Electrician, chemical plant.
Age 46, married, grown children. 1 year in job; 16 years at plant. Union member. — 30, 35, 40, 44, 89, 120, 123, 128, 145, 152, 155, 165

Kitty — Industrial painter, university.
Age 22, single, one child. 2½ years in job; 4 years total as painter. Union member. — 29, 32, 62, 78, 79, 98, 118, 144, 151, 161

Laura — Filter cleaner, pharmaceutical plant.
Age 26, single, no children. 6 years at plant; father and brother employed there. Executive board and grievance committee member of union local. COSH group member. — 6, 39, 46, 64, 76, 80, 129, 140, 146, 154, 166, 172, 176

Lee — Stage carpenter, university.
Age 34, single, no children. 5 years in job; previous employment varied. No union. — 9, 81, 115, 131, 156, 172, 177

Les — Furniture restorer, self-employed.
Age 29, married, 1 child. 16 years in trade, beginning as apprentice in small unionized shop in England. Currently works on commission. No union. — 29, 47, 71, 85, 99, 174

Lisa — Laboratory technician, university.
Age 28, single, no children. 2 years in job; previously held other laboratory jobs. No union. — 13, 60, 73, 167

Lyle — Painter, chemical plant.
Age 40, married, 3 children. 2 years in job, 12 years as chemical operator in plant. Executive board member and shop steward of union local. — 114, 123, 135

| | | |
|---|---|---|
| Mark | Physicist, university.<br>Age 21, married, no children. In doctoral program.<br>Parents and siblings are also physicists. | 72, 86, 97,<br>129, 151,<br>157 |
| Mary | Housewife, wife of a railroad conductor.<br>Age 30, married to Don, four children, one died of<br>cancer attributed to husband's work exposure.<br>Worked briefly as laboratory technician. | 32, 63, 93,<br>129, 144,<br>168, 176 |
| Mel | Silk-screen printer, toy factory.<br>Age 20, single, no children. 10 months in job; quit<br>due to adverse health effects; previous employment<br>varied. Union member. | 72 |
| Mike | Photo lab processor, blueprint company.<br>Age 28, single, no children. 3 years in job; previous<br>employment varied. No union; union organizer. | 10, 28, 32,<br>47, 52, 57,<br>86, 94, 101,<br>133, 161,<br>165, 176 |
| Nick | Chemical operator, chemical plant.<br>Age 43, married, 2 children. 15 years in plant as<br>operator and in maintenance; previous employment<br>varied. Executive board member and grievance<br>chairman of union local. | 28, 54, 63,<br>71, 140, 171 |
| Nora | Graphic artist, blueprint company.<br>Age 38, single, no children. 5 years in job; previous<br>employment varied. No union; union organizer. | 35, 87, 102 |
| Pat | Graphic artist, community agency.<br>Age 35, married, 1 child. 2 months in job, 5 years on<br>this type of machinery, in fine arts 15 years total. No<br>union. | 27, 131 |
| Paula | Electron microscopist, pharmaceutical plant.<br>Age 29, married, no children. 3 years in job, 6 years<br>total at plant; previous employment in biomedical<br>research. No union. | 55 |
| Penny | Laboratory technician, health department.<br>Age 28, married, 1 child. 2 years in department;<br>previously held various laboratory and clerical jobs.<br>Former union member. Currently runs small family<br>farm. | 48, 95 |
| Peter | Railroad signal inspector.<br>Age 35, married, 2 children, both of whom suffered<br>birth deformities attributed to work exposure. 2<br>years in job, 12 years total for railroad. Father and<br>brother also work for railroad. Union member. | 8, 33, 116 |
| Rich | Orchard worker.<br>Age 31, married, 1 child. 5 years in job. Currently<br>works as laboratory technician at same institution.<br>No union; union organizer. | 13, 45, 36,<br>60, 75, 78,<br>108 |

| | | |
|---|---|---|
| Rose | Pill coater, pharmaceutical plant. Age 21, married, no children. 4 years in plant. Husband works at plant. Shop steward of union local. | 44, 48, 58, 154, 167 |
| Sally | Services technician, hospital. Age 37, single, no children. 2½ years in job; previously employed in other laboratories and as a high school chemistry teacher. Union member. | 28, 59, 160 |
| Sandy | Rigger, chemical plant. Age 28, married, young child who suffered seizures attributed to work exposure. 2 years in job, previous 3 years as production operator in same plant. Offical of union local. COSH group member. | 25, 64, 92, 94, 105, 123, 164 |
| Sheila | Laboratory technician, research institute. Age 33, single, no children. 2 years in job, previously employed in related fields. No union. Currently health and safety specialist. COSH group member. | 25, 33, 52, 56, 76, 92, 96, 133, 155, 159 |
| Sid | Chemical operator, chemical plant. Age 25, single, no children. 5 years in job, previous employment as laborer. Health and safety committee member of union local. | 47, 89 |
| Steve | Railroad trackman. Age 33, single, no children. 10 years in job; previous employment varied, unemployed 1 year. Shop steward of union local. | 35, 58, 90, 91, 104, 110, 118, 122, 129, 132, 162, 165, 171, 173 |
| Stuart | Mold maker, glass factory. Age 33, married, 2 children. 15 years in job. Secretary and shop steward of union local. | 41, 70, 89, 95, 145, 159, 162, 170 |
| Sue | Laboratory technician, university. Age 28, married, no children. 2 years in job; previous employment varied. No union. | 53, 162 |
| Ted | Welder, chemical plant. Age 38, married, 1 child. 15 years in job. President of union local. | 40, 65, 86, 109, 115, 130, 142, 155, 167, 175 |
| Tom | Print machine operator, university. Age 27, single, no children. 2 years in job; 8 years in trade. Shop steward and secretary of union local. Health and safety activist. | 55, 102, 123, 127, 156 |
| Tony | Dry cleaner. Age 75, married, 3 grown children. 10 years in job, 50 years in trade. No union. | 14, 81, 153 |

Vivian      Laboratory technician, research institute.              31, 61, 109,
            Age 25, single, no children. 2½ years in job, previ-    157
            ously in school. No union. Union organizer. Cur-
            rently works for consumer health advocacy group.

Walter      Pipe fitter, glass factory.                             4, 53, 61,
            Age 45, married, 4 children. 14 years in job, 32 years  65, 114,
            in trade. Safety committee member and vice-presi-       159, 170,
            dent of union local.                                    176

# Appendix 2A
## Workers' Perceptions of Health Effects from Chemical Exposures

| Respondent | Complaints | Suspected Cause |
|---|---|---|
| Arnie | Nosebleed<br>Convulsions | Chrome<br>Nitrogen oxide |
| Ben | Chemical burns<br>Dry, irritated skin<br>Coughing, eye irritation | Lime dust (calcium oxide)<br>Calcium chloride<br>Ammonia |
| Bess | Inflamed eyes | Nitric and HF (hydrofluoric) acids |
| Carol | Taste dulled | Xylene fumes |
| Daniel | Dry skin<br>Dizziness<br>Headaches | PGM, KAP<br>Ketones<br>Methanol |
| David | Skin rash | An aerosol |
| Debbie | Fibroid growth in lung ("Beautician's lung")<br>Nausea | Hair bleach (Chlorine-based)<br>Chemically heated perms |
| Don and Mary | Child who died of cancer | Don's exposure to dioxin in weed killers |
| Dorothy | Skin rash, dizziness, nausea, blurred vision | Coal tar epoxy |
| Elise | Dry skin | Chemicals, unspecified |

| | | |
|---|---|---|
| Ellen | Headaches | Toluene |
| | Chest pains, irritation of skin, nose, throat | PGCH |
| Eve | Numbness, eye irritation | Trichloroethylene |
| | Skin irritation | Weevil wax |
| Frank | Headaches, dizziness | Cyanide |
| Fred | Cancer of the lip | HCl (Hydrogen chloride) |
| | Headaches, loss of consciousness | Cyanide |
| | Skin irritation | Diethyl benzene |
| | Chemical burn | Cyanoacetic acid |
| Gene | Skin irritation | Phosphorus pentachloride |
| | Breathing obstructed | PGCH, monochloroacetic acid |
| | Headache | Cyanide |
| | Loss of consciousness | Darco (activated carbon) |
| Greg | Dizziness, eye irritation | DMSO (Dimethyl sulfoxide) |
| Irv | Dizziness | Resins and solvents |
| Jack | Dry skin | Cidex (glutaraldehyde) |
| James | Skin irritation | Methyl chloroform |
| | Respiratory, throat irritation | Alkaline solution |
| Jerry | Breathing obstructed | Soda ash, smoke from glass molds |
| | Chemical burns, eye irritation | Soda ash, lime (calcium oxide) |
| Jill | Inflamed eyes, respiratory irritation | Formaldehyde |
| Jocelyn | Eye and skin irritation, dizziness | Diethylamino ethanol |
| Joe | Convulsions, headaches, recurring respiratory problems | Phosgene |
| John | Chemical burns | Orbit (acid bath) |
| Ken | Asthma, severe allergic reaction | PGCH |
| Kitty | Skin irritation | Solvent |
| | Dizziness, headache | Lacquer thinner (aliphatic hydrocarbons) |
| Laura | Nosebleeds, skin irritation | Chemical waste in filters |
| Lee | Nausea | Paint |
| | Spontaneous hemorrhaging of skin, headaches | Acetone |
| Les | Chemical burns | Nitric acid |
| | Temporary loss of sight | Benzene-based stripper |
| | Nausea | Stains |

| | | |
|---|---|---|
| Lyle | Headaches, dizziness, disorientation | Epoxy paint, phosgene, ketones |
| | Allergic reaction | EMME (diethyl [ethoxymethylene] malonate) |
| Mark | Dizziness | Acetone |
| Mel | Tightness in chest, dry skin, dizziness | Paint thinner |
| Mike | Chemical burns, dizziness, disorientation, respiratory irritation, breathing obstructed | Film cleaner, fixer, washout (containing sulfuric acid) |
| Pat | Headaches | Burning vinyl fumes |
| | Nausea | Printing inks, fumes |
| Penny | Dizziness, chronic sinus congestion | Ether, dichloromethane, hexane fumes |
| | | Hexane splash |
| Peter | Child with birth defects | Peter's exposure to dioxin in weed killer |
| Rich | Headache, nausea | Systox (demeton) |
| | Chemical burns | $H_2SO_4$ (sulfuric acid) |
| Sandy | Tremors, blurred vision, dizziness, insomnia, drooling, uncontrollable urination, irritability | Mercury |
| | Infant suffered seizures | Sandy's exposure to mercury |
| Sheila | Nausea, pneumonitis | Pesticides |
| | Headaches, disorientation | Solvents |
| Sid | Skin irritation, chemical burns | Methanol, alcohol, phthalic anhydride |
| | Headaches | Ketones |
| Steve | Suspected liver damage | Dioxin from weed killers |
| Stuart | Dizziness | Paint |
| Ted | Breathing obstructed | Acetylene welding |
| | Burns, punctured eardrum | Chemical explosion |
| Vivian | Headaches | Ozone from copy machine |
| | Skin irritation | Chlorox |
| Walter | Dizziness, nausea | Titanium tetrachloride |
| | Asthma | Sulfur |

# Appendix 2B
## Major Recognized Health Effects
## of Substances Identified by Respondents[1]

**Acetone:** (a ketone solvent) Central nervous system (CNS) depression;[2] eye, skin, mucous membrane irritation[3] at high concentrations or repeated exposure.

**Acetylene:** CNS depression and asphyxiation at high concentrations. Impairs mental alertness, muscular coordination. Possibly causes ventricular fibrillation.

**Activated carbon:** (Darco) No well-demonstrated health hazards.

**Aliphatic Hydrocarbons:** (in lacquer thinner) Petroleum distillates causing eye, skin, mucous membrane irritation, CNS depression, asphyxiation at high concentrations.

**Ammonia:** Eye, skin, and respiratory tract irritation, causing corrosive burns, bronchitis, pulmonary edema at high concentrations.

**Benzene:** Eye, skin, respiratory tract irritation; CNS depression; adverse effects on bone marrow. Acute exposure to high concentrations may cause ventricular fibrillation. Associated with leukemia.

**Calcium chloride:** Toxic effects due to chlorine content (see Chlorine).

**Calcium oxide:** (Lime) Eye, respiratory tract irritation, bronchitis, and

1. For discussion of the relationship between health effects and exposure levels, see Appendix 3.
2. Central nervous system (CNS) depression produces a range of symptoms depending on type and level of exposure: fatigue, weakness, numbness, headache, vertigo, irregular respiration, nausea, vomiting, convulsions, cardiac arrhythmias, coma.
3. Form and severity of irritation depends on whether exposure is through vapor inhalation or direct contact with liquid or dust.

pneumonia. Possible chronic effects are ulceration or perforation of nasal septum.

**Chlorine:** Eye, skin, mucous membrane irritation; prolonged contact causing dermatitis, dental corrosion. High concentrations may cause tracheobronchitis, pulmonary edema, pneumonia.

**Chlorox:** See hydrogen chloride.

**Chrome:** Skin, respiratory tract irritation; high concentrations or prolonged exposure causing asthmatic bronchitis, nosebleeds, pulmonary edema, possibly perforated nasal septum, dental corrosion. Associated with kidney, lung cancer.

**Coal tar epoxy:** Eye, skin, respiratory tract irritation; causing phototoxic skin reactions and acne, bronchitis, conjunctivitis. Associated with skin, lung, bladder, and kidney cancer.

**Cyanide:** CNS depression; at high concentrations causes metabolic asphyxiation.

**Cyanoacetic acid:** Eye, skin irritation.

**Demeton:** (Systox) Exposure to high concentrations results in respiratory and ocular distress, succeeded by gastrointestinal and muscular difficulties, symptoms of CNS depression, temporary paralysis, drop in blood pressure, and cardiac irregularities. Low concentrations are cumulative.

**Dichloromethane:** (methylene chloride) Eye, skin, respiratory tract irritation; CNS depression. High concentrations may cause liver and/or kidney cancer, pulmonary edema, toxic encephalopathy.

**Diethyl benzene:** Eye, skin, respiratory tract irritation; CNS depression. Prolonged exposure may cause dermatitis, bronchitis.

**Diethylamino ethanol:** Eye, skin, respiratory, and gastrointestinal tract irritation.

**Dimethyl sulfoxide:** (DMSO) Eye, skin gastrointestinal tract irritation, causing dermatitis, urticaria, nausea, vomiting, abdominal pain. Possible mutagen, teratogen.

**Dioxin:** Contaminant of 2,4,5-Trichlorophenoxyacetic acid (herbicide). Eye, skin, gastrointestinal tract irritation, causing dermatitis, nausea, vomiting, abdominal pain. Associated with liver damage; suspected mutagen, teratogen, carcinogen.

**Epoxy:** Sometimes contains amine curing agents causing severe eye, skin, respiratory tract irritation, sensitization and asthmatic symptoms.

**Ether:** (Ethyl ether) Eye, skin, respiratory tract irritation. CNS depression at high concentrations. Chronic exposure may cause anorexia, exhaustion, psychological disturbances.

**Formaldehyde:** Eye, skin, respiratory tract irritation. Acute exposure may cause pulmonary edema, pneumonitis. Suspected carcinogen.

**Glutaraldehyde:** (Cidex) Eye, skin, respiratory tract irritation; prolonged exposure causes dermatitis.

**Hexane:** Eye, skin, respiratory tract irritation; CNS depression; asphyxiation at high concentrations. Chronic exposure may cause peripheral neuropathy.

**Hydrofluoric acid:** (hydrogen fluoride) Eye, skin, mucous membrane irritation. Prolonged exposure to low concentrations may cause nasal congestion, bronchitis. High concentrations may cause fever, cyanosis, pulmonary edema.

**Hydrogen chloride:** Eye, skin, respiratory tract irritation. Repeated exposure may cause dermatitis, visual impairment, dental corrosion. High concentrations may cause laryngitis, bronchitis, pulmonary edema.

**KAP:** No information available.

**Ketones:** Eye, skin, mucous membrane irritation; CNS depression. Repeated exposure may cause peripheral neuropathy.

**Mercury:** Acute exposure at high concentrations causes bronchitis, pneumonitis, digestive disturbances, renal damage. Chronic exposure leads to neurologic and psychic disturbances: anorexia; weight loss; stomatitis and excessive salivation; muscular tremors, weakness, loss of coordination; insomnia, irritability.

**Methanol:** (Methyl alcohol) High concentrations may cause optic neuropathy and metabolic acidosis, resulting in impaired vision, headache, nausea, shortness of breath. Prolonged exposure causes dermatitis, digestive disturbances.

**Methyl chloroform:** (1,1,1-Trichloroethane) Eye, skin, respiratory tract irritation; CNS depression. Severe exposure may cause kidney, liver damage. Suspected carcinogen.

**Monochloroacetic acid:** (Chloroacetic acid) Eye, skin, mucous membrane irritation. Systemic effects unknown.

**Nitric acid:** Eye, skin, respiratory tract irritation. Prolonged exposure causes dermatitis, dental corrosion. High concentrations may cause pneumonitis, pulmonary edema.

**Nitrogen oxides:** Eye, mucous membrane irritation; CNS depression. High concentrations may cause pulmonary irritation and methemoglobinemia; chronic exposure may cause emphysema.

**Orbit:** (an acid bath) Unidentified trade name product.

**Ozone:** Mucous membrane, respiratory tract irritation. High concentrations may cause pulmonary edema, headache, burning eyes, changes in visual acuity. Repeated exposure may cause damage to chromosomal structures.

**PGCH:** No information available.

**PGM:** No information available.

**Phosgene:**    Eye, respiratory tract irritation. Chronic exposure to high concentrations may result in pneumonitis, emphysema, cardiac arrest.

**Phosphorus pentachloride:**    Eye, respiratory tract irritation, high concentrations causing pulmonary edema, bronchitis. Repeated exposure causes dermatitis.

**Phthalic anhydride:**    Eye, skin, respiratory tract irritation and sensitization.

**Soda ash:**    (sodium carbonate) Eye, skin, respiratory, gastrointestinal tract irritation.

**Sulfur:**    Eye, skin, mucous membrane irritation at high concentrations.

**Sulfuric acid:**    Eye, skin, respiratory tract irritation; high concentrations causing bronchitis, pulmonary edema, eye and skin burns. Prolonged exposure causes dermatitis, conjunctivitis, stomatitis, dental corrosion, tracheobronchitis.

**Titanium tetrachloride:**    Eye, skin, respiratory tract irritation.

**Toluene:**    Eye, skin, respiratory tract irritation; CNS depression. Long-term exposure may result in liver and kidney damage.

**Trichloroethylene:**    Eye, skin, respiratory, and gastrointestinal tract irritation; CNS depression. Exposure to high concentrations may cause ventricular fibrillation, cardiac arrhythmias, liver and kidney damage, peripheral neuropathy. Suspected carcinogen.

**Xylene:**    Eye, skin, mucous membrane irritation. Exposure to high concentrations may cause CNS depression, temporary renal and/or hepatic impairment, pulmonary edema.

Sources
Proctor, Nick H., and James P. Hughes. *Chemical Hazards of the Workplace.* Philadelphia: J. B. Lippincott Co., 1978.
Sax, N. Irving. *Dangerous Properties of Industrial Materials.* 5th ed. New York: Litton Educational Publishing, Inc., 1979.
Sittig, Marshall. *Handbook of Toxic and Hazardous Chemicals.* Park Ridge, N.J.: Noyes Publications, 1981.
U.S. National Institute for Occupational Safety and Health. *Occupational Diseases: A Guide to Their Recognition.* Rev. ed. Washington, D.C.: DHEW (NIOSH) Publication no. 77-181, 1977.
———. *Registry of Toxic Effects of Chemical Substances.* Vols. 1–2, 1980 ed. Washington, D.C.: DHHS (NIOSH) Publication no. 81-116, 1982.

# Appendix 3
## Toxicological Concepts[1]

Throughout this book we have referred to the difficulties of documenting the actual health effects of exposure to toxic substances. In this Appendix we expand on these difficulties and explain some of the concepts employed by industrial toxicologists as they seek to understand the effects of exposure to the substances and physical agents found in the workplace.

Toxicity and Exposure

The toxicity of a substance or physical agent refers to the quality and degree of its potential adverse effects. A toxic substance does not necessarily constitute a hazard unless people are exposed to it. Toxicologists classify hazardous substances according to their degree of toxicity and the probable human reaction to exposure under varying conditions.

People can be exposed to toxic substances through several routes. They can inhale a substance, absorb it through the skin or mucous membranes, or ingest it. Their exposure to a substance may be short term or long term. Short-term exposure to a massive dose of a toxic substance may have very different effects than low-dose exposures over many years, even though the total amount of exposure is the same.

Defining the Effects of Exposure

Once exposed, individuals may experience an acute (immediate) response, or a chronic effect which develops slowly over a long period of

1. For a much more complete description of these concepts, see references in the Bibliographic Essay, especially for Chapter 2.

time. Short- and long-term exposures may cause either or both types of response. Moreover, these effects can occur at the site of exposure (localized) or, when a substance is absorbed, the effects can occur at some site other than the contact point. They also may occur only after a very long latency period following exposure. Most toxic substances are associated with a variety of health problems; conversely, the same problem may be due to a wide range of substances.

Some substances are more likely to affect particular sites: trichloroethylene affects the skin, dioxane the liver, and organic dusts the lungs. Some tend to affect body systems: organic solvents can damage the central nervous system, carbon monoxide damages the cardiovascular system, while ethylene oxide is a respiratory tract irritant. Because many of these problems are indistinguishable from "ordinary" diseases, it is difficult to identify their relationship to chemical exposure. Certain substances such as lead, some pesticides, and several drugs such as methotrexate affect the reproductive functions either by interfering with conception (sterility), creating changes in genetic material (mutagenicity), or affecting the fetus in the womb, causing birth defects (teratogenicity).

The carcinogenetic effect of certain substances—their role as cancer-causing agents—is well established. As of 1982, the U.S. Department of Health and Human Services has identified some 88 chemicals as carcinogenetic. Some of these substances cause very specific cancers (benzidine has been directly correlated with bladder cancer); others cause a variety of cancers (the vinyl chloride monomer is associated with cancer of the liver, brain, and respiratory system).

Dose-Response Relationships

The concept of a dose-response relationship is fundamental to understanding and regulating the effects of exposure to toxic substances. To determine the degree of acute toxicity, scientists assume that there is a level below which exposures carry minimal risk. This is called the threshold level. This concept, however, is not applied to carcinogenic substances, where any level of exposure is considered hazardous.

All substances are evaluated by a "dose-response curve" that represents the changes in the level of effect that occur with changes in the level of exposure to the substance. Each toxic substance has a characteristic dose-response curve. This concept of a dose-response relationship suggests that as exposure increases so too does the likelihood of an adverse effect. Similarly, it is assumed that harm can be mitigated by reducing the level of exposure.

Quantifying this relationship, however, is difficult, and much of the controversy over regulatory standards focuses on defining threshold levels and dose-response curves for specific chemicals. The route of exposure, form of the substance, temperature, humidity, physiologic condition of the person exposed, and the synergistic effects of exposure in conjunction with other substances all may bear on the relationships between exposure level and effects on the health of an individual. These variables leave considerable room for disagreement as interested parties dispute the appropriate criteria for determining the probability of risk, establishing the severity of health effects, and deciding the nature of regulation.

Dose-response curves and causal relationships between substances and ill health can be evaluated through experimental tests on animals, observations of human experience, or short-term bioassays. Toxicologists use animal tests to classify the effects of various toxic substances. They determine a dose-response curve by administering a range of doses to experimental animals under a variety of conditions. Acute toxicity is measured by estimating the median lethal dose ($LD_{50}$), that is, the dose of a substance sufficient to kill half of a group of animals subject to short-term exposure.

To evaluate chronic effects, toxicologists undertake long-term studies giving various dose levels to test animals over time and over at least two generations. The test animals are autopsied to locate affected organs and the cause of death. These tests form the basis of dose-response curves. Such tests can cost over $500,000 per substance. Since it would be prohibitively expensive to use hundreds of animals in each test, the researcher will administer high doses of a chemical in order to increase the likelihood of obtaining statistically valid results. Lower dose effects are obtained by extrapolations from high dose effects. Because different species may vary in their responses, the validity of using animal tests to predict human responses is viewed with skepticism.

Direct evidence of harmful effects on humans can be derived from human experimentation and epidemiological studies. The first method is limited by ethical and legal considerations. The latter is complicated by the difficulty of systematically determining the substances that are present in diverse situations, the precise levels of exposure, the routes of exposure, the synergistic effects of multiple exposures, and the variations in individual sensitivity and personal habit. For these reasons epidemiological studies are seldom able to establish definitive causal relationships, except in special cases where exposure is limited to a definable population and the health effects are specific (e.g., the rare liver cancer caused by exposure to vinyl chloride monomers). However, epidemiological studies have been useful in identifying likely associations.

Short-term bioassays are designed to detect mutagenic substances by testing their effect on reproduction or change in the genetic material of a cell. Their value lies in the observed relationship between mutagenicity and carcinogenicity: over 90 percent of those substances found to be mutagenic are also carcinogenic. While bioassays do not identify all of the known mutagens and carcinogens, they are an essential screening tool necessary to identify priorities for further testing.

Establishing Standards

Since its formation in 1970, the Occupational Safety and Health Administration has been responsible for setting federal standards on workplace exposures. Individual states with agreements to operate their own safety and health programs may set more stringent standards. The government role in defining acceptable levels of workplace exposure is complicated by the uncertainties inherent in the field of toxics, where results at best provide only probabilities. Uncertainties allow interested parties to force their priorities and concerns about costs into the standard-setting process. "How safe is safe enough" is a political, not a technical, decision.

Many of the federal standards are based on recommendations made by a private organization, the American Conference of Governmental Industrial Hygienists (ACGIH), which sets guidelines called threshold limit values (TLVs) for airborne concentrations of substances. They determine TLVs on the basis of the level at which "nearly all workers may be repeatedly exposed day after day without adverse effect." They recognize, however, that "a small percentage of workers may experience discomfort from some substances at concentrations at or below the threshold limit; a smaller percentage may be affected more seriously by aggravations of a pre-existing condition or by development of an occupational illness."

When OSHA was first established the agency adopted many of these TLVs as the basis for government standards. Subsequently, changes in standards and new standards have followed extensive rule-making proceedings. Under the OSHAct these must be set at the level "which most adequately assures, to the extent feasible, on the basis of the best available evidence, that no employee will suffer material impairment of health or functional capacity even if such employee has regular exposure to the hazard dealt with by such standard for the period of his working life." Typically, in regulatory proceedings, industry uses the uncertainties in the data and the questions of feasibility to argue for less stringent standards, while labor and the public health community argue for a maximum degree of protection.

In light of the uncertainty of toxicological and medical information, the appropriate standards for worker protection remain subject to continued dispute. In any case, the adequacy of any standard lies not only in its margin of safety but in its actual enforcement in the workplace.

# Appendix 4
## Sample Material Safety Data Sheet

## U.S. DEPARTMENT OF LABOR
### Occupational Safety and Health Administration

# MATERIAL SAFETY DATA SHEET

Required under USDL Safety and Health Regulations for Ship Repairing,
Shipbuilding, and Shipbreaking (29 CFR 1915, 1916, 1917)

Form Approved
OMB No. 44-R1387

## SECTION I

| MANUFACTURER'S NAME | EMERGENCY TELEPHONE NO. |
|---|---|
| ADDRESS *(Number, Street, City, State, and ZIP Code)* | |
| CHEMICAL NAME AND SYNONYMS    Benzene* | TRADE NAME AND SYNONYMS    Benzol |
| CHEMICAL FAMILY    Organic | FORMULA    $C_6H_6$ |

## SECTION II - HAZARDOUS INGREDIENTS

| PAINTS, PRESERVATIVES, & SOLVENTS | % | TLV (Units) | ALLOYS AND METALLIC COATINGS | % | TLV (Units) |
|---|---|---|---|---|---|
| PIGMENTS | | | BASE METAL | | |
| CATALYST | | | ALLOYS | | |
| VEHICLE | | | METALLIC COATINGS | | |
| SOLVENTS | | | FILLER METAL PLUS COATING OR CORE FLUX | | |
| ADDITIVES | | | OTHERS | | |
| OTHERS | | | | | |

| HAZARDOUS MIXTURES OF OTHER LIQUIDS, SOLIDS, OR GASES | % | TLV (Units) |
|---|---|---|
| | | |
| | | |
| | | |

## SECTION III - PHYSICAL DATA

| | | | |
|---|---|---|---|
| BOILING POINT (°F.) | 176° | SPECIFIC GRAVITY ($H_2O$=1) | 0.88 |
| VAPOR PRESSURE (mm Hg.) @ 26.1° C. | 100mm | PERCENT, VOLATILE BY VOLUME (%) | 100 |
| VAPOR DENSITY (AIR=1) | 2.77 | EVAPORATION RATE (____ =1) | |
| SOLUBILITY IN WATER | sl. Sol | | |
| APPEARANCE AND ODOR | Clear colorless liquid, characteristic odor. | | |

## SECTION IV - FIRE AND EXPLOSION HAZARD DATA

| FLASH POINT (Method used) | FLAMMABLE LIMITS | Lel | Uel |
|---|---|---|---|
| 12° F    (Closed Cup) | | 1.3% | 7.1% |
| EXTINGUISHING MEDIA    Foam is best.  $CO_2$ and dry chemical | | | |
| SPECIAL FIRE FIGHTING PROCEDURES    Wear self-contained breathing apparatus. | | | |
| UNUSUAL FIRE AND EXPLOSION HAZARDS    When fighting fire use necessary protective equipment and breathing apparatus. | | | |

*See note in Section No. IX - Special Precautions
(Continued on reverse side)

PAGE (1)                                                                      Form OSHA-20

## SECTION V - HEALTH HAZARD DATA

THRESHOLD LIMIT VALUE    25 mg/$M^3$ Air.    $LD_{50}$ ORAL (RAT) = 3400mg/Kg.

EFFECTS OF OVEREXPOSURE
Dizziness, headache, breathlessness, and excitement followed by mental confusion and hysterical symptoms.

EMERGENCY AND FIRST AID PROCEDURES
1.) Remove from exposure 2.)   Remove contaminated clothing  3.)   If in eyes, flush with water for 15 minutes.  Call physician.  4.)   Internally - induce vomiting, call doctor.  Do not give anything by mouth if unconscious  5.)   If inhalation - lie down, keep warm, call physician.

## SECTION VI - REACTIVITY DATA

| STABILITY | | CONDITIONS TO AVOID |
|---|---|---|
| | UNSTABLE | |
| | STABLE    X | |

INCOMPATABILITY *(Materials to avoid)*    Oxidizing materials

HAZARDOUS DECOMPOSITION PRODUCTS
Yes - All types of products in case of fire.

| HAZARDOUS POLYMERIZATION | | CONDITIONS TO AVOID |
|---|---|---|
| | MAY OCCUR | |
| | WILL NOT OCCUR    X | |

## SECTION VII - SPILL OR LEAK PROCEDURES

STEPS TO BE TAKEN IN CASE MATERIAL IS RELEASED OR SPILLED
Small quantities - absorb on paper.  Evaporate in hood.  Wear proper safety equipment.

WASTE DISPOSAL METHOD
Atomize into an incinerator

## SECTION VIII - SPECIAL PROTECTION INFORMATION

RESPIRATORY PROTECTION *(Specify type)*
Handle in hood or use self-contained breathing apparatus.

| VENTILATION | LOCAL EXHAUST | SPECIAL |
| | MECHANICAL *(General)* X (Sparkproof Fans) | OTHER |

| PROTECTIVE GLOVES | EYE PROTECTION |
| Synthetic rubber | Cup type or rubber framed goggles. |

OTHER PROTECTIVE EQUIPMENT

## SECTION IX - SPECIAL PRECAUTIONS

PRECAUTIONS TO BE TAKEN IN HANDLING AND STORING
Outside or detached storage is preferable.

OTHER PRECAUTIONS
Inside storage should be in a standard flammable liquids storage room or
cabinet.  NOTE:  This item is listed as a cancer-suspect agent by OSHA as of
6/2/80.

**PAGE (2)**
GPO 9 30-540

**Form OSHA-20**
Rev. May 72

# Bibliographic Essay

This essay includes material that we have found especially useful as background reading and references in our research. It is by no means a comprehensive bibliography, but a selection of some key books, articles, and technical resources on the subjects covered in each chapter.

**Introduction: The Dangerous Trades**

An early discussion of the problems of occupational health can be found in the autobiography of Alice Hamilton, *Exploring the Dangerous Trades* (Boston: Little, Brown, 1943).

A comprehensive bibliography on the field of risk analysis is in Vincent Covello and Mark Abernathy, "Risk Analysis and Technological Hazards: A Policy-related Bibliography," in C. Whipple et al., *Technological Risk Assessment* (Netherlands: Sijthoff and Nordhoff, 1983). Useful books and articles are as follows: Chauncey Starr, "Social Benefit vs. Technological Risk," *Science* 165 (September 19, 1969), 1232–1238; William Lowrance, *Of Acceptable Risk* (Los Altos, Calif.: Kaufman, 1976); William Rowe, *An Anatomy of Risk* (New York: Wiley, 1977); R. W. Kates, *Risk Assessment of Environmental Hazards* (New York: Wiley, 1979); Baruch Fischoff, Paul Slovic, and Sarah Lichtenstein, "Which Risks Are Acceptable," *Environment* 21 (May 1979), 17–38; and Baruch Fischoff, Sarah Lichtenstein, Paul Slovic, S. Derby, and R. Keeney, *Acceptable Risk* (New York:

Cambridge University Press, 1981); R. C. Schwing and W. A. Albers, eds., *Societal Risk Assessment: How Safe Is Safe Enough?* (New York: Plenum Press, 1980); and M. Dierkes et al., eds., *Technological Risk: Its Perception and Handling in the European Community* (Cambridge, Mass: Oelgeschlager, Gunn and Hain, 1980).

Books and articles focusing specifically on risk as a social concept are: Mary Douglas and Aaron Wildavsky, *Risk and Culture* (Berkeley: University of California, 1982); and Michael Thompson and Aaron Wildavsky, "A Proposal to Create a Cultural Theory of Risk," in *The Risk Analysis Controversy,* ed. H. Kunreuther and E. Ley (Berlin: Springer-Verlag, 1982).

A number of books deal with a wide spectrum of issues concerning occupational health. These include: Nicholas Ashford, *Crisis in the Workplace* (Cambridge: MIT Press, 1976); and Ray Davidson, *Peril on the Job: A Study of Hazards in the Chemical Industries* (Washington, D.C.: Public Affairs Press, 1970).

There are relatively few surveys of workers' attitudes toward risk. A major study is the Michigan survey on the quality of employment: Robert P. Quinn and Graham L. Staines, *The 1977 Quality of Employment Survey; Descriptive Statistics, with Comparison Data from the 1969–70 and the 1972–73 Surveys* (Ann Arbor: Institute for Social Research, University of Michigan, 1979). For analyses of this survey, see Richard Frankel and W. Curtiss Priest *Health, Safety and the Worker,* Department of Labor Report ASPER/Con-78/0103/A, National Technical Information Service, Springfield, Va., 1979. Also, Shell Oil commissioned a survey: *An Analysis of Public and Worker Attitudes toward Carcinogens and Cancer Risks,* Cambridge Reports, April 1978. A survey approaching questions similar to those raised in this study is Julia Green Brody, "Workers' Perceptions of Long Term Occupational Health Hazards" (Ph.D. diss., Department of Psychology), University of Texas at Austin, May 1983.

There are many reference books with statistical information on occupational injuries, on the toxic effects of chemical substances, and on the chemical industry. These include: Bureau of Labor Statistics, *Occupational Injuries and Illnesses in the U.S. by Industry* (USGPO, Bulletin 2078, August 1980); Environmental Protection Agency, *Toxic Substances Control Act Chemical Substance Inventory* (EPA Office of Toxic Substances, 1979); USHHS, PHS, CDC, NIOSH, *1980 Registry of Toxic Effects of Chemical Substances,* DHHS (NIOSH) Publication no. 81-116 (Washington, D.C.: USGPO, 1982); Department of Labor, *Interim Report to Congress on Occupational Disease* (Washington, D.C.: U.S. Department of Labor, 1980); "Facts and Figures for the Chemical Industry," *Chemical and Engineering News* (June 8, 1981), 30–72; and James Wei,

*The Structure of the Chemical Processing Industries: Function and Economics* (New York: McGraw-Hill, 1979).

For bibliographic material on occupational health see: USDHHS, NIOSH, *NIOSH Publications Catalog,* DHHS (NIOSH Publication no. 80-126 (Washington, D.C.: USGPO, 1980); and U.S. Department of Labor, Office of the Assistant Secretary for Administration and Management, *Occupational Safety and Health: A Bibliography* (Washington, D.C.: USDOL, 1979).

### Chapter 1: Working with Chemicals

For general background on the risks of working with chemicals, see: Jeanne Stellman and Susan Daum, *Work Is Dangerous to Your Health* (New York: Vintage Books, 1973); and USDHEW, PHS, DCD, NIOSH, *National Occupational Hazard Survey,* DHEW (NIOSH) Publication no. 78-114 (Washington, D.C.: USGPO, 1977).

On work, and in particular on how technological change and managerial strategies affect the work process, see: Andrew Zimbalist, ed., *Case Studies on the Labor Process* (New York: Monthly Review Press, 1979); Michael Buroway, "The Anthropology of Industrial Work," *Annual Review of Anthropology* 8 (1979), 231–266, and *Manufacturing Consent: Changes in the Labor Process under Monopoly Capitalism* (Chicago: University of Chicago Press, 1979); and Studs Terkel, *Working* (New York: Pantheon Books, 1974). For the classic history of the American working class, see Philip Foner, *History of the Labor Movement in the United States,* 5 vols. (New York: International Publishers, 1942). For occupational health issues pertaining to women, see Jeanne M. Stellman, *Women's Work, Women's Health: Myths and Realities* (New York: Pantheon Books, 1977).

### Chapter 2: Illnesses and Complaints

Some standard technical reference books on industrial toxicology and occupational diseases are: F. A. Patty, *Patty's Industrial Hygiene and Toxicology,* ed. George D. Clayton and Florence E. Clayton, 3 vols., 3d rev. ed. (New York: Wiley, 1978); Marcus Key et al., *Occupational Diseases: A Guide to Their Recognition,* DHEW (NIOSH) Publication no. 77-181 (Washington D.C.: USGPO, 1977); Stanley Kusnetz and Marilyn K. Hutchison, eds., *A Guide to the Work-relatedness of Disease,* rev. ed., USDHEW (NIOSH) Publication no. 79-116 (Washington, D.C.: USGPO, 1979); and Alexander McRae and Leslie Whelchel, eds., *Toxic Substances Control Source Book* (Germantown, Md.: Aspen Systems Corp., 1978); *TLV's, Threshold Limit Values for Chemical Substances and Physical Agents in the Work Environment with Intended Changes for 1982,* American Conference of Governmental Industrial Hygienists, Cincinnati, Ohio (1982); *Documentation of the Threshold Limit Values*, 4th ed., American Conference of Governmental Industrial Hygienists, Cincinnati, Ohio (1980).

A number of books and articles specifically focus on cancer in the workplace. These include: Devra Lee Davis, K. Bridbord, and M. Schneiderman, "Estimating Cancer Causes: Problems in Methodology, Production and Trends," *Banbury Report 9: Quantification of Occupational Cancer* (Cold Springs Harbor Laboratory, New York, 1981), and "Cancer in the Workplace," *Environment* 23, no. 6 (July–August 1981), 25–37; see, also, Samuel S. Epstein, *The Politics of Cancer* (New York: Anchor Press, 1979). A critique of Epstein's book is Richard Peto, "Distorting the Epidemiology of Cancer: The Need for a More Balanced View," *Nature* 284 (27 March 1980), 297–300. An industrial perspective focusing on environmental sources of cancer is developed in Merril Eisenbud, "Environmental Causes of Cancer," *Environment* 20, no. 8 (1978), 6–16; and Merril Eisenbud, *Environment, Technology and Health* (New York: New York University Press, 1978).

For the neurological effects of exposure to toxic substances, see: Barry Johnson and W. K. Anger, "Behavioral Toxicology," in *Environmental and Occupational Medicine*, ed. Willan Rom (Boston: Little, Brown, 1982).

For the effects of chemical exposure on reproduction, see: Council on Environmental Quality, *Chemical Hazards to Human Reproduction* (Washington, D.C.: USGPO, 1981); Coalition for the Reproductive Rights of Workers, *Reproductive Hazards in the Workplace: A Resource Guide* (Washington, D.C.: n.d.); Wendy Williams, "Firing the Woman to Protect the Fetus," *Georgetown Law Journal* 69 (1981), 641–704.

Some industrial statistics on illness appear in the National Safety Council: *Accident Facts—1979 Edition*, National Safety Council, 444 N. Michigan Avenue, Chicago, IL 60611; The National Paints and Coatings Association, *A General Mortality Study of Production Workers in the Paints and Coatings Manufacturing Industry*, 1500 Rhode Island Avenue, N.W., Washington, D.C. 20005; and B. W. Karrh, M.D., S. Pell, M. T. O'Berg, "Cancer Epidemiologic Surveillance in the DuPont Company," *Journal of Occupational Medicine* 20, no. 11 (November 1978), 725.

On workers accuracy in self-reporting illness, see U.S. Department of Labor, *An Interim Report to Congress on Occupational Disease* (Washington, D.C.: USGPO, 1980).

### Chapter 3: Anxieties and Fears

The anxiety related to illness has been a subject of psychological analysis in Robert Lifton's *The Broken Connection* (New York: Simon and Schuster, 1979); literary description in Susan Sontag's *Illness as Met-*

*aphor* (New York: Farrar, Straus and Giroux, 1978); and anthropological discussion in Arthur Kleinman, "The Meaning Context of Illness and Care," in *Sciences and Cultures: Sociology of the Sciences,* ed. E. Mendelsohn and Y. Elkana, vol. 5 (Dordrecht: Reidel, 1981), 161–176.

The social factors contributing to anxiety at work are discussed in Richard Kazis and Richard Grossman, *Fear at Work* (New York: Pilgrim Press, 1982).

### Chapter 4: What's to Blame?

The issue of blame for the problems of workplace risk is addressed in: Council on Economic Priorities, *Occupational Safety and Health in the Chemical Industry* (New York: Council on Economic Priorities, 1981); Daniel Berman, *Death on the Job* (New York: Monthly Review Press, 1978); Rachel Scott, *Muscle and Blood* (New York: Dutton, 1974); and Audrey Freedman, *Industry Response to Health Risk* (New York: Conference Board, 1981). For a government analysis, see Peter J. Sheridan, "What's Causing Mysterious Illnesses?—NIOSH Seeks Answers," *Occupational Hazards* (April 1980), pp. 66 ff.

### Chapter 5: Protection on the Job

Technical material on respirators and other protective devices appears in: OSHA, *A Worker's Guide to the Use of Respirators* (Washington, D.C.: n.d.); NIOSH/OSHA, *Pocket Guide to Chemical Hazards,* DHEW (NIOSH) Publication no. 78-210 (Washington, D.C.: USGPO, 1978); NIOSH/OSHA, *Occupational Health Guidelines for Chemical Hazards,* DHHS (NIOSH) Publication no. 81-123 (Washington, D.C.: USGPO, 1981); American National Standards Institutes, *Practices for Respiratory Protection,* Z88.2 (New York, 1980); L. R. Birkner, *Respiratory Protection, A Manual and Guideline* (Akron, Ohio: American Industrial Hygiene Association, 1980). For a catalog of available protection devices, see the periodical *Industrial Hygiene News.*

See, also: U.S. Department of Labor, *Protecting People at Work: A Reader in Occupational Safety and Health* (Washington, D.C.: USDOL, 1980); and Nick H. Proctor and James P. Hughes, *Chemical Hazards of the Work Place* (Philadelphia: Lippincott, 1978).

### Chapter 6: Adaptations

Several studies deal with adaptation to dangerous jobs. See: John S. Fitzpatrick, "Adapting to Danger: A Participant Observation Study of an Underground Mine," *Sociology of Work and Occupations* 7, no. 2 (May 1980), 131–158; David P. McCaffrey, *OSHA and the Politics of Health Regulation* (New York: Plenum Press, 1982), especially chap. 3. Jack Haas, "Learning Real Feelings: A Study of High Steel Ironworkers'

Reactions to Fear and Danger," *Sociology of Work and Occupations* 4, no. 2 (May 1977), 147–170. Robert Blauner, in *Alienation and Freedom* (Chicago: University of Chicago Press, 1964), and Kenneth Lasson, in *The Workers* (New York: Bantam Books, 1971), deal with the issue of alienation and discontent in different work settings. Albert O. Hirschman, in *Exit, Voice, and Loyalty: Responses to Decline in Firms, Organizations, and States* (Cambridge, Mass.: Harvard University Press, 1970), examines alternative strategies to deal with dissatisfaction at work.

### Chapter 7: Activism

Among the books on union activism are: Alice Lynd and Staughton Lynd, *Rank and File: Personal Histories by Working-Class Organizers*, 2d ed. (Princeton, N.J.: Princeton University Press, 1981); Pat Cayo Sexton, *The New Nightingales* (New York: Enquiry Press, 1982); Solomon Barkin, ed., *Worker Militancy and Its Consequences, 1965–1975: New Directions in Western Industrial Relations* (New York: Praeger, 1975); Thomas Kochan et al., "Collective Bargaining and the Quality of Work: The Views of Local Union Activists," in Industrial Relations Research Association Series, *Proceedings of the 27th Annual Winter Meeting, 1974*, pp. 150–162; John McDermott, *The Crisis in the Working Class and Some Arguments for a New Labor Movement* (Boston: South End Press, 1980).

Several newsletters and periodicals regularly contain articles on shop floor activism. See, e.g.: *Labor Notes* (Labor Education and Research Project, Detroit, Mich.); *American Labor* (American Labor Education Center, Washington, D.C.); *In These Times* (Transnational Institute, Chicago, Ill.); *Union Democracy Review* (Association for Union Democracy, Brooklyn, N.Y.).

### Chapter 8: If There's a Hazard . . .

On the environmental policies of the chemical industry, see *Chemecology,* the monthly journal of the Chemical Manufacturers Association. For a corporate view of industrial policy, see F. I. Dupont deNemours, *Occupational Safety and Health: A DuPont Company View* (April 1980) (available from Public Affairs Department, Wilmington, Del. 19898); and Chemical Manufacturers Association, *Worker Safety in the Chemical Industry . . . What We're Doing about It* (Washington, D.C., 1980). On the role of safety committees, see Thomas A. Kochan, Lee Dyer, and David B. Lipsky, *The Effectiveness of Union-Management Safety and Health Committees* (Kalamazoo, Mich.: W. E. Upjohn Institute for Employment Research, 1977).

For material on the union role in occupational health, see: Thomas Kochan, *Collective Bargaining and Industrial Relations* (Homewood, Ill.: Richard D. Irwin, Inc. 1980); Lawrence S. Bacow, *Bargaining for Job*

*Safety and Health,* (Cambridge, Mass: M.I.T. Press, 1980); Carl Gersung, *Work Hazards and Industrial Conflict* (Kingston, R.I.: New England Press, 1981).

### Chapter 9: If I Call OSHA . . .

The key legislation covering occupational health is the Occupational Safety and Health Act, 29 USC 651 et seq. A publication reporting on relevant government activity in the area of occupational health is: Bureau of Affairs, *OSH Reporter* (Washington, D.C.: BNA, weekly). See, also, *Occupational Health and Safety Letter,* Environnews, Inc., 1097 National Press Bldg. Washington, D.C. 20045 (semiweekly)

Questions of governmental and scientific responsibility in regulating under conditions of uncertainty are discussed in David Bazelon, "Risk and Responsibility," *Science* 205 (1979), 277–281.

Two contrasting views of OSHA's role appear in Council on Economic Priorities, *Occupational Safety and Health in the Chemical Industry* (New York: LEP, 1981), and H. R. Northoup, R. L. Rowan, and C. R. Perry, *The Impact of OSHA: A Study of the Effects of the Occupational Safety and Health Act on Three Key Industries—Aerospace, Chemicals, and Textiles* (Philadelphia: Wharton School, University of Pennsylvania, 1978).

### Chapter 10: If I'm Sick . . .

Considering the growing importance of occupational medicine, there is little material on the role of the company doctor and corporate health policies. See the *Journal of Occupational Medicine;* also, the New York Academy of Medicine, ed., "Ethical Issues in Occupational Medicine," *Bulletin of the New York Academy of Medicine* 54, no. 8 (September 1978); and George Annas, "Legal Aspects of Medical Confidentiality in the Occupational Setting," *Journal of Occupational Medicine* 18, no. 8 (August 1976), 537–540. For relationships between medical services and compensation, see: John Blum, "Corporate Liability for In-House Medical Malpractice," *St. Louis University Law Review* 22, no. 3 (1978), 451–453; and Frank Goldsmith and L. E. Kerr, *Occupational Safety and Health Administration: The Prevention and Control of Work Related Hazards* (New York: Human Science Press, 1982). For a union perspective on occupational medicine, see Thomas Mancuso, *Helping the Working Wounded* (Washington, D.C.: International Association of Machinists, 1976).

On compensation, see: Peter S. Barth, with H. Allan Hunt, *Workers' Compensation and Work-related Illnesses and Diseases* (Cambridge, Mass: M.I.T. Press, 1980); and LaVerne C. Tinsley, "Workers Compensation in 1980: Summary of Major Enactments," *Monthly Labor Review*

3 (1981). For a general critique of compensation law, see "The Benefit and the Doubt," *Health/PAC* 12, no. 4 (March/April 1981), 18–24. For the inequities and inadequacies of state compensation laws, see *The Report of the National Commission on State Workmen's Compensation Laws,* AD/7816/UG/A6, Washington, D.C., July 1972.

### Chapter 11: Knowing the Risks

On the right to know legislation and its implementation, see: Nancy Kim, *Right to Know Bill—Implementation and Work Plan* (Albany, N.Y.: Department of Health, November 1980); Frank Goldsmith, "The Right to Know and the Right to Refuse Hazardous Work," *Consumer Health Perspectives* 6 (January 1980); Mary Melville, "Risks on the Job: The Worker's Right to Know," *Environment* 23, no. 9 (November 1981), 12–20, 41–45; and Elihu D. Richter, "The Worker's Right to Know: Obstacles, Ambiguities and Loopholes," *Journal of Health Politics, Policy and Law* 6, no. 2 (Summer 1981), 339–346.

On COSH groups, see Daniel Berman, "Organizing for Job Safety," *Science for the People* (July/August 1980), 11–15.

Federal policy on the communication of hazards is discussed in: U.S. Department of Labor, Occupational Safety and Health Administration, "Hazards Identification: Notice of Proposed Rulemaking and Public Hearing," 46*FR* 4412 (16 January 1981), Proposal withdrawn 46 *FR* 12214, and "Hazard Communication; Proposed Rule and Public Hearing Announcement," 47*FR* 12092 (19 March 1982). On labeling, see P. Slovic, B. Fischoff, and S. Lichtenstein, "Informing People about Risk," in *Product Labelling and Health Risks,* ed. L. Morris et al., Banbury Report 6, Cold Springs Harbor Laboratory, New York, 1980.

For a listing of union publications created under OSHA's New Directions program, see U.S. Department of Labor, OSHA, *A Resource Guide to Worker Education Materials in Occupational Safety and Health* (Washington, D.C.: USDOL, 1982).

A key article on the obstacle of trade secrecy is Thomas McGarity and Sidney Shapiro, "The Trade Secret Status of Health and Safety Testing Information," *Harvard Law Review* 93, no. 5 (March 1980), 837–889.

### Chapter 12: Controlling the Risks

On the question of the "right to health" and its relationship to work, see U.S. Department of Labor, Occupational Safety and Health Administration, Committee on Public Information in the Prevention of Occupational Cancer, Division of Medical Sciences, Assembly of Life Sciences, National Research Council, National Academy of Sciences,

*Informing Workers and Employers about Occupational Cancer* (Washington, D.C.: USGPO, 1978).

The issues involved in paying workers in hazardous jobs extra money for taking risks are developed in Julie Graham and Don M. Shakow, "Risk and Reward Pay for Workers," *Environment* 23, no. 8 (October 1981), 14–21.

There is an extensive literature on workers' control and industrial democracy dealing with both theoretical and practical issues. See Sven Forssman, *The New Work Environment* (Stockholm: Swedish Institute of Occupational Health, 1975); Barry Maley, Dexter Dunphy, Bill Ford, *Industrial Democracy and Worker Participation* (Adelaide, Australia: Unit for Industrial Democracy, South Australian Department of Labour and Industry, 1979); M. Bertrand, "Sur la Securité du Travail: 15 Morts par Jour," *Economie et Politique* (Paris) 270 (January 1977), 55–59.

For a bibliography on worker participation, see: Maryse Gaudier, *Workers' Participation in Management, Selected Bibliography, 1977–79,* (Geneva: International Institute for Labor Studies, 1981). An extensive collection of reprints and documents is at the Documentation Center of the Participation and Labor-managed Systems Program of Cornell University, Ithaca, N.Y. 14853.

For specific applications to occupational safety and health, see: Steven Deutsch, "Extending Workplace Democracy: Struggles to Come in Job Safety and Health," *Labor Studies Journal* 6, no. 1 (Spring 1981), 124–132; Joan Brown, "Occupational Health and Safety: The Importance of Worker Participation," *Labour Gazette* 78 (April 1978), 123–128; Noronha Filho Gerson, "The Self-Management Model and Its Relevance to the Conditions of the Working Environment in Yugoslavia: An Exploratory Evaluation" (thesis, Public Health, Johns Hopkins University, September 1977); Swedish Trade Union Confederation and Swedish Central Organization of Salaried Employees, "Report on the Workplace Environment" (paper delivered at the European Trade Union Conference, Geneva, 28 February–1 March 1975).

Recent research linking job control to stress and cardiovascular disease is reported in Robert Karasek, et al., "Job Decision Latitude, Job Demands, and Cardiovascular Disease: A Prospective Study of Swedish Men," *American Journal of Public Health* 71, no. 7 (July 1981), 694–705.

For information on joint committees as a form of participation, see Nicholas Ashford and Sally Owen, *Labor-managed Safety and Health Committees: Developing a Structure for Decision Making and for Future Policy Research* (Cambridge, Mass: Center for Policy Alternatives, M.I.T., 1979).

The issues of power and control that pervade this analysis are discussed in similar terms in Steven Lukes, *Power: A Radical View* (Tiptree, Essex: Anchor Press, 1974); and John Gaventa, *Power and Powerlessness: Quiescence and Rebellion in an Appalachian Valley* (Urbana: University of Illinois Press, 1980).